More Praise for *Precision Medicine, AI and the Science of Personalized Healthcare*

"I have had the great pleasure and joy of working with Jim Wallace for many years. He is the living embodiment of the 'Do good and do well' innovation ethic so much needed in healthcare."
—Regina E. Herzlinger, Nancy R. McPherson
Professor of Business Administration,
Harvard Business School

"Clear, well-researched, and clinically relevant—this book shows how pharmacogenomics and medication optimization can deliver on the promise of safer, more effective treatments."
—Yoona Kim, CEO, *Arine*

"*Precision Medicine, AI and the Science of Personalized Healthcare* offers a compelling vision for the future of patient-centered care. James Wallace masterfully synthesizes today's most important clinical innovations— from pharmacogenomics and digital biomarkers to AI-driven diagnostics and predictive analytics—into a readable, deeply human narrative. As a Chief Medical Officer, I'm particularly struck by the actionable clarity he brings to topics like genotype-guided prescribing, dynamic risk modeling, and the integration of wearables with remote monitoring. Each chapter demonstrates how real-world application of precision medicine can simultaneously improve outcomes, reduce inefficiencies, and empower patients. This book doesn't just highlight the science; it reveals the future of care delivery—personalized, proactive, and profoundly data-informed. A must-read for clinicians, executives, and health policy leaders who believe that the path forward is one of precision, compassion, and intelligent design. Wallace has provided the roadmap. Now it's the job of clinician executives to implement!"

—Steven Goldberg, MD, Chief Medical Officer,
HealthTrackRx

"Jim Wallace's book brings precision medicine to life through powerful, real-world examples. I was especially struck by the stories of patients—Michael and Jennifer, Emily, Amy, Susan—they're a practical, relatable look at how AI, wearables, and genomics can transform care. *Precision Medicine* isn't just theory—it's a guide to what's possible today."

—John D. Couris, President and CEO,
Florida Health Science Center—Tampa General Hospital

Precision Medicine, AI and the Science of Personalized Healthcare

Precision Medicine, AI and the Science of Personalized Healthcare

James Wallace

Foreword by Prof. Regina Herzlinger,
Harvard Business School

WILEY

To Kristie, for believing

Epigraph

The river flows freely, but it cannot go anywhere else.
—J.W.

Epigraph

Contents at a Glance

Contents

Foreword

For more than two decades, I've argued that empowering the patient is the most sustainable path to transforming healthcare. Consumer-driven healthcare is not a slogan, it is a moral and economic imperative. And it is finally within reach—thanks to the convergence of technologies and ideas identified in this book.

In *Precision Medicine, AI and the Science of Personalized Healthcare*, James Wallace doesn't just echo that message. He advances it.

Drawing on years of experience as a healthcare strategist and innovator and our shared work as a senior researcher for the *Innovating in Healthcare* course at Harvard Business School, Jim delivers a clear, compelling, and necessary vision for where healthcare must go. Through richly drawn patient stories, rigorous analysis, and forthright critique, he makes the case that precision medicine, grounded in data, technology, and consumer autonomy, is the most hopeful and actionable strategy for reform.

My own research—spanning dozens of Harvard Business School cases and academic publications across the full continuum of care—has focused on how markets, innovation, and consumer choice can drive quality up and costs down. Jim shares that orientation, but what sets this book apart is its ability to translate theory into reality. He shows us what consumer-driven care *looks like* in the age of wearables, genomics, Big Data, and AI, and what stands in its way.

This is not a book of empty optimism, but a call to arms. Jim understands that the forces creating friction against change—strong incumbents, misaligned payment models, and regulatory inertia—do not move easily. But he also understands that empowered patients, equipped with the right tools and knowledge, are a powerful force to suppress.

If you are an investor, policymaker, clinician, or healthcare entrepreneur, this book will sharpen your strategy. If you are a patient—and we all are—it will give you reason to believe that the future of medicine can be more personal, more effective, and more just.

Jim Wallace is helping lead the transformation. Read this work, and then join the movement.

—**Regina E. Herzlinger**
Nancy R. McPherson Professor of
Business Administration
Harvard Business School
Author, *Consumer-Driven Health Care* and
Innovating in Healthcare

Preface

We live in the age Alvin Toffler foresaw in *Future Shock*—a time when information moves faster than institutions, and individuals must adapt faster than the systems meant to support them. Nowhere is this dislocation more profound than in healthcare. The explosion of data, from wearable sensors, genetic testing, electronic health records, and behavioral signals, has outpaced our ability to make it useful. This book is about how we catch up. More precisely, it is about how we integrate these fragments into a coherent, actionable whole.

Artificial intelligence and Big Data are not merely tools for efficiency; they are catalysts for understanding. When harnessed correctly, they offer the promise of medicine that isn't just smart, but personal. But information without integration is just noise. Precision medicine begins when we combine data with insight, and insight with individual relevance.

This book is not about technology for its own sake. It is about the recentering of care around the patient—not in the aspirational language of slogans, but in the actionable reality of one person, one body, one decision at a time. Healthcare has always been, and always will be, self-driven. Doctors are not the architects of health; they are its facilitators, its advisors, and sometimes its engineers. But the person at the center of the system is not the provider—it is the patient.

Our system today remains structured around episodic, reactive events: visits, prescriptions, procedures. But these are only waypoints in a much longer journey. Just as medicine advanced from heroism to empiricism to evidence-based care, we now approach the next transition: from fragmented episodes to integrated, longitudinal care. Proactive care is not only more effective, it is inevitable.

Unfortunately, our current healthcare systems are built for delay, not foresight. Traditional medicine, especially in the United States, is a fragile, negative-sum game of trial and error. It's costly and inefficient, and too often it leaves the patient feeling disempowered. Value-based care, for all its admirable intentions, remains hamstrung, trying to encourage the wrong person (the physician) to engage the wrong way (with money) at the wrong time (after the fact). Care providers want to help people be healthier, and patients want to avoid illness and injury. Only precision medicine offers an anti-fragile, positive-sum path forward—one that improves as it scales, adapts as it learns, and strengthens through variation.

This transformation also requires a new mental model for how care is delivered. The artificial divide between "primary" and "specialty" care no longer serves the complexity of modern medicine. Instead, we need a longitudinal model based on phases of care: preventive (pre-condition), diagnostic (inception), interventional, and long-term management. These are the new segments of a personalized care continuum—described in Chapter 12—that better match the rhythm of real life and the trajectory of real disease.

The goal of this book is not only to inform but also to inspire. We are standing at the edge of a remarkable future—one where medicine embraces the unique contours of each individual life. Like the river in the epigraph, personalized medicine moves freely, drawing power from its flow. And yet, when viewed from the right distance, we see it was never arbitrary. It goes where it must.

Let's follow it.

—James Wallace
Boston, Massachusetts
June 2025

Introduction: The Dawn of a New Era

The basis of traditional healthcare is the notion that we are all alike: alike because we share common illnesses, conditions, and issues that can be treated effectively by the marvels and discoveries of science.

The reality of healthcare is that we are all different in our DNA, our lifestyles, and our social and physical environments. The earlier we embrace these truths in our journeys of healing, the more we share in successful outcomes.

That difference between traditional medicine, which treats us all the same with widely varying outcomes, and precision medicine, which treats us all differently with tailored and more successful outcomes, is the story of this book.

The Promise of Precision Medicine

Consider this: every year, nearly 50% of patients with common medical conditions do not respond to the first medication prescribed to them.[1]

The causes of this are both clinical, such as non-tolerability, and self-created (not completing the prescription), but the statistic is not just a number; it is a real-world example of how the one-size-fits-all approach in medicine often leads to months, sometimes years, of trial and error, unnecessary side effects, and escalating expense.

Imagine, instead, if each prescription was matched to the individual's unique genetic makeup, lifestyle, and environment—transforming the trial-and-error process into a precise, tailored solution from the start. This is the promise of precision medicine, and it is why the need for a shift in the traditional approach has never been more urgent.

Precision medicine is a transformative approach to healthcare that tailors medical treatment to the individual characteristics of each patient. Unlike traditional medicine, which typically relies on standardized treatments designed for the "average" patient, precision medicine uses genetic information, advanced analytics, and lifestyle factors to develop personalized treatment plans.

Foundational to this approach is genomics, which provides insights into a patient's unique genetic makeup and likely response to medications or interventions. Advanced data analysis then allows healthcare providers to interpret this genetic information alongside vast clinical data, identifying patterns that can guide more effective, personalized care.

By focusing on the specific needs of each individual, precision medicine aims not only to improve health outcomes but also to reduce unnecessary side effects and healthcare costs—marking a significant shift from a one-size-fits-all model.

Precision medicine holds the potential to significantly improve health outcomes by matching treatments to the unique profiles of individual patients, allowing for interventions that are both more effective and less likely to cause harmful side effects. For example, in cancer treatment, precision medicine can identify genetic mutations driving tumor growth, enabling doctors to prescribe targeted therapies that attack the cancer cells without damaging healthy tissue—a strategy that has shown promising results in improving survival rates and quality of life.

In another scenario, consider two patients with heart disease: one might have a genetic predisposition to respond better to a particular drug, whereas the other might require lifestyle adjustments and alternative medications due to different metabolic factors. Precision medicine helps identify these distinctions early, tailoring care to avoid the prolonged trials of traditional treatments.

In more complex cases, like rare genetic disorders, precision medicine can help identify specific gene mutations, allowing for personalized interventions that would be impossible in a one-size-fits-all model. By focusing on the underlying causes rather than symptoms alone, precision medicine has the potential to transform the treatment landscape for both common and rare conditions.

Precision medicine offers promising economic benefits, too, by reducing healthcare costs through more targeted, effective treatments. Traditional medicine often involves a trial-and-error approach, especially for chronic or complex conditions, which can result in multiple failed treatments, hospitalizations, and prolonged medical expenses. Precision medicine addresses this by identifying the most effective treatment at the outset, minimizing unnecessary interventions and side effects that require costly follow-up care. For instance, genetic testing can reveal whether a patient is likely to respond to a specific medication, helping avoid the costs of ineffective drugs.

These improvements extend to broader economic impacts. For insurance companies and other payors, precision medicine can produce substantial cost savings by reducing claims associated with repeated hospital visits, complications, and long-term treatment failures. For patients, personalized care means fewer unexpected medical bills and a lower burden of side effects, which can also translate into fewer missed workdays and a higher quality of life. Ultimately, by increasing treatment efficacy and streamlining healthcare delivery, precision medicine not only reduces individual costs but also eases financial pressures across the entire system, creating a more sustainable healthcare model.

Precision medicine empowers patients by giving them access to personalized information about their unique health profiles, allowing for more informed decision-making and active involvement in their care. With insights from genetic testing and advanced data analysis, patients can understand how specific treatments may impact them based on their own biology, lifestyle, and environment. This knowledge fosters greater autonomy, as patients are no longer limited to generic options but can choose from treatments tailored to their needs and preferences. For example, a patient with a genetic predisposition to a particular side effect can work with their doctor to select safer alternatives, taking control of their healthcare journey. This level of engagement transforms patients into active participants rather than passive recipients, making them collaborators in achieving the best possible outcomes and ultimately enhancing their satisfaction and trust in the healthcare process.

Although precision medicine holds immense promise, several barriers hinder its widespread adoption. High costs are a significant challenge; genetic testing and advanced diagnostics often require expensive technologies, making them inaccessible to many patients without substantial insurance coverage. Additionally, access to precision medicine varies widely, with rural and low-income communities frequently lacking the infrastructure or specialist resources needed to implement these personalized approaches. These differences contribute to growing gaps, as patients in underserved areas remain reliant on traditional, one-size-fits-all treatments that may not meet their unique needs.

Technological limitations also present hurdles. Integrating genetic and clinical data into meaningful, usable insights demands advanced data systems and skilled practitioners, both of which are unevenly distributed across healthcare settings. Privacy concerns further complicate data sharing as current frameworks struggle to protect sensitive genetic information effectively.

Addressing these challenges requires policy approaches that make precision medicine more affordable and accessible. Increased funding, insurance reform, and targeted investment in underserved areas could help access, enabling broader patient benefit from personalized treatments. By tackling these barriers, we can work toward a healthcare system where precision medicine's benefits are within reach broadly.

The future of healthcare is deeply personal, driven by the transformative power of precision medicine to address individual needs and improve lives. By shifting from generic treatments to tailored approaches, precision medicine has the potential to revolutionize outcomes, reduce costs, and empower patients in unprecedented ways. This book brings that promise to life through human stories—patients who have faced difficult diagnoses and found hope in personalized care. Through their journeys, readers will see how precision medicine isn't just advancing science; it's reshaping lives, one unique individual at a time, illuminating the path toward a more personalized healthcare future.

We are standing at the edge of a seismic shift in healthcare. The traditional model of medicine—based on a broad, one-size-fits-all approach—is rapidly being replaced by far more personal, precise, and impactful precision medicine. The new frontier represents a revolution, promising not only better outcomes but also the potential to drastically reduce costs. In this book, we will explore the transformative power of precision medicine and its potential to reshape our understanding of healthcare, treatment, and cost, paving the way for more effective, personalized, and accessible healthcare.

Patient-Focused Approach to Healthcare

Imagine a patient with chronic pain who has been prescribed a series of standard medications over months, each with limited success and debilitating side effects. Although traditional healthcare often follows protocols based on what works for the average patient, this approach overlooks the individual's unique experience and needs—leaving the patient feeling frustrated, unheard, and still in pain.

A patient-focused approach, by contrast, starts with a deep understanding of the person behind the symptoms. Instead of merely treating the disease, healthcare providers engage with the patient's lifestyle, goals, and personal concerns. This approach emphasizes personalized care, focusing on solutions that address the patient as a whole rather than a diagnosis alone. Through this lens, precision medicine becomes a powerful tool for crafting treatments that truly fit the patient, making healthcare more compassionate, effective, and respectful of each person's unique journey.

A patient-focused approach to healthcare is about prioritizing the well-being, lifestyle, values, and unique circumstances of each individual, rather than relying on standardized, one-size-fits-all treatments. In this model, healthcare is tailored to fit the patient's life, not the other way around, creating a supportive environment where medical decisions are collaborative and focused on what matters most to the person receiving care.

This approach aligns with emerging trends in patient-centered care and shared decision-making, where patients and providers work together to select treatments that best meet the patient's health goals, values, and preferences. By acknowledging the unique aspects of each individual's circumstances, patient-focused healthcare fosters a more effective system that recognizes individual needs and empowers patients to take an active role in their care. It's a philosophy that moves beyond treating symptoms and addresses the full scope of a person's health, enhancing satisfaction and health outcomes alike.

A patient-focused approach to healthcare leads to better health outcomes by addressing the unique factors influencing each patient's life and health. By considering an individual's personal circumstances, lifestyle, and genetic uniqueness, this approach helps healthcare providers craft more effective, personalized care plans that patients can realistically integrate into their daily lives. For example, a diabetes patient with a busy work schedule might benefit from a tailored medication and meal plan that considers their limited time, making it easier to adhere to their treatment and avoid complications.

Personalized care also boosts patient satisfaction, as patients participate more in the decision-making process. This, in turn, fosters greater collaboration with their healthcare team and higher adherence to prescribed treatments, which is crucial for chronic conditions where long-term management is key. Similarly, a cancer patient receiving treatment tailored to their genetic profile may experience fewer side effects and a faster recovery, improving both their health and overall quality of life.

By treating people as unique individuals rather than cases, a patient-focused approach creates a more effective healthcare experience, empowering patients and ultimately improving their physical and emotional well-being.

At the same time, precision medicine empowers patients on a clinical level by providing tailored, evidence-based insights that enhance their understanding of their own health and treatment options. By leveraging genetic data, biomarkers, and other advanced diagnostics, precision medicine allows healthcare providers to present patients with specific, data-driven information about how different treatments might impact them personally. This scientific clarity enables patients to make informed choices based on the expected efficacy and potential side effects tailored to their unique biological profile.

For instance, a cancer patient might receive a genetic analysis indicating a particular drug's likelihood of success in targeting their tumor type, allowing them to bypass standard, less effective treatments and reduce trial and error. Similarly, for patients with chronic conditions like cardiovascular disease, precision medicine can reveal metabolic markers that help identify the most compatible medications, minimizing adverse reactions and optimizing outcomes.

By focusing on individualized data, precision medicine transforms patients into active participants in their care, empowering them with precise options grounded in clinical science. This approach fosters confidence in treatment decisions and encourages adherence, as patients understand that their plan is based on robust, personalized evidence rather than generalized protocols.

Implementing a patient-focused approach in healthcare faces significant challenges, starting with practical barriers like time constraints and standardized protocols. Many healthcare providers operate under intense time pressure, limiting their ability to thoroughly assess each patient's unique needs, preferences, and circumstances. Standardized treatment protocols, although efficient, often emphasize uniformity over personalization, making it difficult for providers to adjust care based on individual factors without deviating from established guidelines.

Resource limitations further complicate individualized care. Many healthcare settings lack the staff or tools to perform the necessary assessments, data analysis, and follow-up for a truly personalized approach. Additionally, systemic barriers such as cost structures and insurance claims policies often disincentivize customized care. Insurance reimbursement models tend to favor quick, standardized treatments over comprehensive, patient-centered approaches, making it financially challenging for providers to allocate time and resources toward tailored care.

But precision medicine reduces wasted time and resources, so addressing these obstacles will require clinical reforms aimed at prioritizing patient-centered outcomes, as well as insurance models that support personalized care. Shifting to value-based payment systems that reward individualized treatment and outcome improvements can encourage broader adoption, making patient-focused healthcare more accessible and sustainable for both providers and patients.

A patient-focused approach complements precision medicine by emphasizing care tailored to each patient's unique needs, values, and circumstances. Together, they create a healthcare model that not only improves clinical outcomes but also transforms the patient experience, making it more empathetic and responsive. By aligning personalized treatments with whole-person care, this approach fosters a system where patients feel engaged and participate in their healthcare journey. The future of healthcare lies in this blend of scientific precision and engagement, a vision that promises to elevate outcomes, enhance satisfaction, and make healthcare a revolutionary experience for every individual.

Four Drivers: Wearables, Pharmacogenomics, Big Data Analytics, and Artificial Intelligence

Technology and data are rapidly transforming healthcare, paving the way for a new era of precision medicine that can deliver highly personalized treatments. Advances in data collection and analysis allow healthcare providers to go beyond traditional approaches, integrating real-time patient data and genetic insights into customized care plans. At the forefront of this transformation are four key drivers: wearables, pharmacogenomics, Big Data analytics, and artificial intelligence (see Figure I-1).

Each of these key drivers is different, yet all radically personalize health and healthcare. For example, wearables, such as fitness trackers, smartwatches, and smart clothing, or clinically sophisticated devices like glucose monitors, heart-rate bracelets, and smart hearing aids, enable continuous health monitoring, providing insights into a patient's daily metrics. Pharmacogenomics, on the other hand, combining genetic testing with medication therapy optimization, tailors drug treatments based on genetic

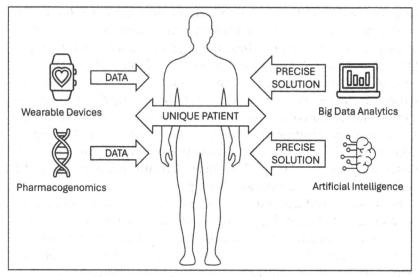

Figure I-1: Relationship of four drivers to patient-focused care

profiles, enhancing effectiveness. Big Data analytics, assessing individual characteristics against patterns identified across vast health datasets, guide more accurate diagnoses earlier in an individual's care path; and artificial intelligence accelerates the interpretation of this wealth of information, helping clinicians make more informed decisions along the way. Together, these pillars support the future of precision medicine, where treatment becomes increasingly predictive, preventive, and personalized.

Wearables

Wearable devices are transforming healthcare by enabling continuous health monitoring, offering real-time data on vital signs such as heart rate, physical activity, sleep patterns, and more. Unlike periodic check-ups, which rely on the patient and provider to engage in real time (and reliably over time), wearables provide a stream of information, creating a detailed picture of a patient's health over a number of days, weeks, or months. This data can be used to proactively manage health, detect significant issues early, and support ongoing management of chronic conditions. For instance, devices with electrocardiogram capabilities can alert patients to irregular heart rhythms, potentially warning them of conditions like atrial fibrillation before they become severe.

This immediate feedback empowers patients to engage actively in their health, making adjustments in response to alerts or trends. For those managing conditions such as diabetes or hypertension, wearables can track blood sugar levels or blood pressure, alerting patients and healthcare providers to potential problems before they escalate. By encouraging early intervention and helping patients understand the impact of their lifestyle on their health, wearables foster a proactive, engaged approach to healthcare. This shift supports better health outcomes and can reduce the need for emergency interventions, as patients are empowered with real-time insights to take charge of their well-being.

Pharmacogenomics

Pharmacogenomics is the study of how a person's genetic profile influences their response to medications, and it plays a pivotal role in the development of personalized treatments. By understanding specific genetic markers, pharmacogenomics allows healthcare providers to select medications that are most likely to be effective for an individual, minimizing the risks of adverse drug reactions and enhancing treatment efficacy. For example, certain genes affect how quickly or slowly the body metabolizes drugs. In cases where a patient metabolizes a drug too rapidly, the medication may not reach therapeutic levels, whereas a slower metabolism could increase the risk of toxicity.

From cardiovascular disease to cancer and from pain management to diabetes, an individual's unique genetic makeup affects not only the likelihood of occurrence but also the effectiveness of treatment. One powerful example is in behavioral health. For patients dealing with depression, pharmacogenomic testing can reveal genetic markers that indicate which antidepressants are likely to work best, sparing them the typical trial-and-error process that can delay relief and result in side effects. By matching the right drug to the genetic profile from the start, pharmacogenomics can streamline treatment, improve patient outcomes, and reduce healthcare costs associated with ineffective medication. This targeted approach not only optimizes care but also offers a glimpse into the future of truly personalized medicine, where treatments are as unique as the individuals receiving them.

Big Data Analytics

From the microscopic level of genomic data, innovation is also occurring at the population level with "Big Data." Big Data is revolutionizing healthcare by aggregating vast amounts of information, including patient records, genetic data, and lifestyle factors, to uncover trends and patterns that were previously invisible or difficult to discern. This ability to integrate and analyze diverse datasets enables healthcare providers to move beyond reactive care to proactive and predictive approaches. Big Data analytics helps identify health risks, manage chronic conditions, and refine treatment strategies, making care more precise and effective.

For instance, analyzing large-scale patient data can reveal population-level risk factors for chronic diseases such as diabetes and cardiovascular disease. Healthcare systems can then use this insight to develop targeted prevention programs for at-risk groups. On an individual level, Big Data can predict a patient's likelihood of developing a condition based on genetic predisposition and lifestyle, allowing for early intervention.

In treatment, Big Data aids in identifying which therapies work best for specific patient subgroups. For example, by examining outcomes across thousands of cases, analytics can highlight which cancer treatments are most effective for patients with particular genetic mutations. By transforming complex information into actionable insights, Big Data combined with sophisticated analytics is an essential driver of precision medicine, enhancing both individual outcomes and the efficiency of healthcare systems.

Artificial Intelligence

Artificial intelligence (AI) in healthcare refers to the use of advanced algorithms and machine learning to analyze complex medical data, enhancing diagnostic accuracy, predictive capabilities, and clinical decision-making. AI can process vast datasets—such as imaging scans, genetic information, and electronic health records—far faster and more accurately than human practitioners, uncovering patterns that may otherwise go unnoticed. This capability is particularly valuable in early disease detection and the creation of personalized treatment plans.

For example, AI-driven imaging analysis has shown remarkable success in detecting cancer at early stages. By analyzing thousands of imaging scans, AI systems can identify subtle signs of tumors with high precision, even at stages where human detection is challenging. Another application is in personalized medicine, where AI algorithms evaluate a patient's unique data, including genetic markers and health history, to recommend specific treatments most likely to succeed.

These predictive and diagnostic functions not only improve patient outcomes by catching diseases early and tailoring treatments but also support healthcare providers by delivering insights that streamline the decision-making process. As AI evolves, it stands to become an increasingly valuable tool, making precision medicine more accessible, efficient, and effective.

The future of healthcare combines the impact of precision medicine with the personalization of patient-focused care.

Importantly, the four drivers—wearables, pharmacogenomics, Big Data, and AI—significant as they are alone, also work in synergy to create a powerful, integrated system that enhances precision medicine. Wearables continuously monitor a patient's health data, providing real-time insights into vital signs or chronic conditions, populating a Big Data platform where it is aggregated with other patient records, lifestyle factors, and genetic information to identify trends and refine patient profiles. Sophisticated AI then analyzes these complex datasets, detecting potential health risks and offering predictive insights that would be difficult if not impossible to discern manually.

Simultaneously, pharmacogenomics employs genetic data to recommend personalized medication choices, reducing trial and error and minimizing adverse reactions. For example, a patient with hypertension might use a wearable device to monitor their blood pressure daily, and AI interprets this data, alerting healthcare providers to irregularities. Analyzing Big Data helps refine this information, and pharmacogenomic testing tailors medication precisely to the patient's genetic makeup. Together, these drivers transform a previously disparate care approach into a seamless, personalized system that improves outcomes, reduces risks, and empowers patients to manage their health more effectively.

The combined power of wearables, pharmacogenomics, Big Data, and AI is paving the way for a transformative future in precision medicine. These technologies enable personalized treatments, early intervention, and tailored patient care, enhancing outcomes and reducing unnecessary treatments. Together, they empower both patients and providers with insights and tools to make healthcare more responsive and effective. As these drivers continue to evolve, they lay a robust foundation for a healthcare system that's more precise, proactive, and patient-centered, marking a shift toward a future where medical care is as unique as the individuals it serves.

Our Guides and Their Stories

Storytelling has the unique power to humanize complex ideas, turning abstract concepts into relatable, real-life experiences. In this book, we bring the transformative potential of precision medicine to life through the journeys of five individuals: Michael, Jennifer, Emily, Amy, and Susan. Their stories illuminate both the promise and challenges of this groundbreaking approach. From a middle-aged couple with kids in school who address their struggles with weight gain to a cancer patient benefiting from AI-guided treatments, their experiences show how precision medicine can change lives.

These narratives invite readers to connect emotionally with their struggles and triumphs, offering a window into the human side of medical innovation. By following these individuals, we not only explore the science behind precision medicine but also feel its impact—bringing hope, progress, and the ongoing quest for accessible, personalized care into practical focus.

Let's meet our guides.

Michael and Jennifer: Fast Food, Busy Lives

Michael and Jennifer, a married couple in their early 40s, have spent years battling the slow march of weight gain, a struggle that has strained their health, relationships, and daily lives. Despite countless diets and exercise plans, they found nothing sustainable—until wearable fitness technology introduced them to a new, personalized approach. Equipped with devices that tracked their activity, heart rates, and even sleep patterns, Michael and Jennifer gain real-time insights into their habits and health metrics. For the first time, they can see how small lifestyle adjustments, like tailored meal plans and consistent movement, impact their progress.

Their journey isn't just physical; it is deeply emotional. By holding each other accountable and celebrating shared milestones, even among their hectic work lives and parenting, they rediscover a sense of partnership and support. Over time, they experience dramatic transformations—not only in their weight but also in their energy, confidence, and quality of life.

However, Michael and Jennifer's story highlights troubling obstacles. The cost of wearable devices and the limits of insurance coverage for associated programs are significant hurdles, raising questions about accessibility for anyone facing similar challenges. Their journey underscores how wearables can revolutionize health management, as well as the significance of raising awareness of these tools and their usefulness for all who can benefit from them.

Emily's Unrest

Emily, a 30-year-old professional, has spent years with varying anxiety and depression, often feeling trapped in a system that relies on impersonal trial-and-error approaches to treatment. Each new medication brings the same cycle: side effects, sometimes debilitating, with little improvement, leaving her frustrated and hopeless. Changing health plans brings a new doctor: a young physician who introduces her to pharmacogenomics, a breakthrough in personalized medicine. By analyzing her genetic profile, Emily discovers why certain medications have failed her and which ones are likely to work best.

With this newfound insight, Emily's treatment plan shifts to a medication tailored to her unique genetic makeup. Within weeks, she begins to feel a stability and clarity she hasn't experienced in years. For the first time, Emily has the tools to manage her mental health effectively and regain her confidence.

Emily's journey isn't without challenges, though. Accessing pharmacogenomic testing presents financial and logistical hurdles, and the emotional toll of navigating mental health conditions often leaves her feeling isolated. Emily's story highlights not only the transformative potential of precision medicine in mental health but also the obstacles that must be addressed to ensure that patients can benefit from these life-changing advancements.

The Joy and Pain of Being Amy

Amy, a 50-year-old wife and mother, has spent years battling chronic pain that disrupted her daily life and strained her relationships. Traditional medicine offered little more than temporary relief and vague diagnoses, leaving her feeling unheard and defeated. Her breakthrough came when her healthcare team turned to Big Data analytics. By aggregating and analyzing her medical history, genetic information, and lifestyle data alongside thousands of similar cases, they identified patterns that pointed to an underlying condition previously overlooked.

This data-driven approach enables Amy's team to craft a targeted pain management strategy, incorporating a combination of personalized therapies and lifestyle adjustments. For the first time in years, Amy experiences relief, allowing her to rebuild relationships and return to the activities she loves.

Amy's story underscores the transformative potential of data-driven healthcare, turning ambiguous symptoms into actionable insights that dramatically improve outcomes. Amy's journey also exposes barriers, like inefficiencies of traditional healthcare systems and limited access to advanced techniques required for precision medicine. Despite these challenges, Amy's case illustrates how harnessing Big Data can not only alleviate suffering but also restore hope and functionality to patients long left in the shadow of vague diagnoses.

Susan's Battle with Cancer

Susan, a 64-year-old mother and grandmother, was devastated when diagnosed with late-stage cancer, a prognosis that left her with limited options under traditional treatment protocols. Feeling helpless, she found a glimmer of hope in an innovative approach powered by AI. AI allows an in-depth analysis of Susan's unique cancer profile, incorporating genetic mutations, tumor characteristics, and treatment history into a comprehensive model. This analysis identifies a targeted therapy that aligns perfectly with her cancer's specific biology—an option that had not been considered before.

The treatment plan brings renewed hope and a tangible improvement in Susan's quality of life, turning despair into determination. Susan's journey highlights the emotional rollercoaster of confronting cancer and the revolutionary potential of AI-driven precision medicine to deliver personalized, life-saving solutions.

Susan's story also sheds light on the complexities of adopting AI in oncology. High costs, limited access to advanced AI tools, and the challenges of robust data-sharing infrastructure remain significant barriers. Despite these hurdles, Susan's experience showcases the transformative impact AI can have, offering tailored solutions that can extend lives and improve outcomes. Her story underscores the urgency of integrating these technologies into healthcare systems to make their life-saving potential more accessible.

Precision Medicine Transforms Lives

The stories of Michael, Jennifer, Emily, Amy, and Susan illuminate the universal impact of precision medicine, demonstrating its ability to address diverse challenges across different demographics and medical conditions. From combating obesity and managing chronic pain to overcoming mental health struggles and battling late-stage cancer, their experiences reveal how personalized care can transform lives. These narratives also reflect broader healthcare realities: the inefficiencies of one-size-fits-all approaches, systemic barriers to access, and the emotional toll of navigating traditional medical systems.

Yet their journeys also offer hope and inspiration. Precision medicine, driven by cutting-edge technologies and patient-focused care, shows what's possible when treatment is tailored to the individual. Their triumphs underscore the potential for a more empathetic and effective healthcare system where science and compassion intersect to improve outcomes for all. These stories set the stage for deeper exploration, illustrating the promise of a future where healthcare is truly personal.

Throughout the book, Michael, Jennifer, Emily, Amy, and Susan will serve as touchstones, guiding readers through the transformative world of precision medicine. Their stories—rich with struggles and triumphs—bring the science to life, making complex concepts relatable and deeply human. By connecting with their journeys, readers can envision how precision medicine might impact their own lives or those of their loved ones, offering hope for more personalized and effective care. As these guides navigate challenges and breakthroughs, they illuminate the path forward, demonstrating that the future of healthcare isn't just innovative—it's profoundly personal.

Note

1. Brown, Marie, and Bussell, Jennifer, "Medication Adherence: WHO Cares?" *Mayo Clinic Proceedings* 86, no. 4 (2011): 304–314.

Michael and Jennifer: A Family's Fight Against Obesity

Mornings in the Hansen household were a whirlwind of activity, a blend of organized chaos and quiet determination. Michael and Jennifer moved with practiced efficiency, navigating the demands of three energetic kids—Noah, 12, Jacob, 10, and Emma, 8—while juggling their own busy schedules. Fortunate to have Michael's mother Susan living close for the occasional last-minute requirement, the family was a team. Ever the numbers guy, Michael was already mentally sorting through the day's meetings as he poured cereal into bowls. Jennifer, with her signature flair, balanced a mug of coffee while reminding Emma to grab her soccer cleats.

Their marriage was solid, a partnership forged through years of love, laughter, and shared goals. But as they caught glimpses of each other between the chaos—Michael pulling his fingers through his hair in the mirror, Jennifer tying hers back before heading to the salon—both felt a pang of nostalgia for their younger selves. Back then, they were carefree, full of energy, and the idea of a Friday night out didn't feel like a Herculean effort. Life changed, as it does, and although their love remained steady, the subtle shifts couldn't be ignored.

The weight gain had crept up gradually, first with the sleepless nights of parenting, then with the hurried meals and skipped workouts that came with building careers and raising kids. They weren't unhappy, but they missed feeling their best. The thought wasn't always at the forefront, but it lingered—a quiet acknowledgment that their lives, although full, could be better if they felt better.

As Michael grabbed his backpack and Jennifer kissed Emma good-bye, the couple exchanged a quick smile. The day was starting, full and demanding, with an unspoken truth between them: they were in this together.

As the morning bustle faded and the kids began their day at school, the silence settled over Michael and Jennifer. Their lives were busy, even successful, but beneath the surface lay a shared struggle that neither could ignore. The weight they carried wasn't just physical; it was emotional—a quiet but persistent burden that touched every corner of their lives.

Michael felt it as he climbed the stairs to the entrance of the office, his breath coming faster than it used to, and as he glanced at the photos on his desk of the family not so long ago. Jennifer felt it too, in the way her lower back and feet ached after a long day at the salon and how she avoided joining the kids as they gamboled about on returning home. It wasn't just about the numbers on the scale; it was about what those numbers represented—shrinking opportunities, dwindling energy, and dreams they'd put on hold.

Michael and Jennifer's relationship bore the weight as well. Although their marriage remained strong, the little things—like holding hands during a walk or sharing a spontaneous dance in the kitchen—happened less often. The frustration they felt about their own struggles sometimes spilled over, unspoken but felt, creating moments of distance in their otherwise loving partnership.

Most of all, it was the weight of what could be that lingered. They both wanted more—not just for themselves, but for their kids, who looked—reluctantly, but still looked—to them for guidance and inspiration. They dreamed of family hikes, of running alongside Noah at his next track meet, of feeling vibrant and free again. Yet as the days passed, they found themselves wondering how to begin—how to reclaim the energy and joy that seemed just out of reach.

That afternoon, the drive home from the doctor's office was unusually quiet. Michael gripped the steering wheel, his knuckles firm as Jennifer stared out the window, lost in thought. The afternoon sunlight filtered through the windshield, warm and golden, but it felt distant, almost irrelevant. Coming from Michael's annual physical exam, they both heard the doctor's words echoing in their minds: elevated cholesterol, prediabetes, joint strain. A shared reality they'd known, but hearing it out loud felt heavier than expected.

Jennifer finally broke the silence. "He's right, you know. We need to make some changes." Her voice was steady, but there was a flicker of vulnerability in her tone. Michael exhaled, the tension in his shoulders easing as he nodded. "I know. I've known for a while," he admitted, glancing at her apprehensively before returning his focus to the road.

They sat with the weight of it for a moment, letting the truth settle between them. It wasn't about blame or about shame. It was about facing something they could no longer avoid. Michael thought of Noah's track meet the next weekend and how he'd love to run alongside him without gasping for breath. Jennifer thought of chasing Emma across the backyard, laughing without worrying about her sore back and legs.

As they pulled into the driveway, Michael turned off the car but didn't move. Jennifer reached for his hand, their fingers intertwining in a quiet gesture of love. "We've done hard things before," she said softly. Michael smiled faintly, the corners of his mouth lifting just enough. "We can do this, too."

In that moment, there was no grand plan or sudden clarity, just a shared understanding that life wasn't just about numbers or health metrics. It was about reclaiming the life they wanted—together.

Michael: The Silent Fighter

Michael leaned back in his office chair, staring at the spreadsheet on his computer screen. Numbers were his comfort zone, neat and orderly, with clear answers for every question. But life isn't a spreadsheet, and for years, Michael had felt the messy, uncontrollable weight of his own body—and the emotions that came with it—slowing him down.

He wasn't always this way. In his twenties, Michael was active, lean, and full of confidence. Back then, he wouldn't hesitate to join a pickup basketball game or worry about how he looked in a family photo. But marriage, kids, and the responsibilities that followed slowly shifted his priorities. Quick meals became the norm, late-night snacks a coping mechanism for stress, and gym sessions a distant memory. The weight crept on gradually, almost imperceptibly at first. Then, one day, he realized he could no longer ignore it.

Michael hid his disappointment well. At work, he cracked jokes during meetings, keeping the mood light and drawing attention away from himself. With friends, he'd be the reliable one, always ready with a laugh or a story. Even at home, he masked insecurities with humor, deflecting conversations about his health with self-deprecating remarks that felt more safe than honest.

But beneath the surface, Michael struggled. He hated the way his shirt felt tighter around his stomach, the way his neck and shoulders ached after a long day at the office, the way he avoided mirrors because they reflected more than his physical form—they reminded him of the disappointment he felt in himself. He worried about his kids, too. What kind of example was he setting for them? He told himself it wasn't too late to change, but the thought of that was overwhelming.

What stung the most was how his weight affected the small joys in life. He used to love tossing a football with Noah, running alongside him until they both collapsed in laughter. Now, just the idea of running felt daunting. He missed dancing with Jennifer in the kitchen, a spontaneous joy that now felt like a distant memory.

Michael didn't talk about these feelings, not even with Jennifer. He convinced himself she wouldn't understand, even though he knew she was fighting her own battles. Instead, he kept moving forward, focusing on work, on providing for his family, on anything but himself. But the doctor's appointment changed something. For the first time, Michael felt the weight wasn't just physical; it was emotional, and it was holding him back from being the husband, dad, and man he wanted to be.

As he sat in the car in the driveway after work, watching his kids cavort through the house, Michael felt something stir—an inkling of determination. He didn't know how to start or what it would take, but he knew one thing: he had to try. For them. And for himself.

Michael couldn't pinpoint the exact moment his weight began to take over his life, but he could see its effects in every corner of his day. The tendonitis in his elbow had started months ago, a nagging pain that he chalked up to overuse. He told himself it was just part of getting older, but he knew the extra pounds weren't helping. Even the simplest tasks—carrying groceries, lifting a ladder—became cautious calculations, each movement a reminder of how much his body had changed.

The mornings were always the hardest. Michael sat in his car, the engine humming softly, staring at the long line snaking through the McDonald's drive-through. "Why am I here again?" he thought, drumming his fingers on the steering wheel. The answer was obvious: it was easy. Too easy. No chopping vegetables, no packing a lunch the night before—just a quick stop on the way to work.

Still, he couldn't shake the frustration bubbling inside him. It wasn't the weight that got to him. No, it was the familiar, sinking feeling of defeat, the knowledge that he was breaking yet another promise to himself. "I could've made oatmeal this morning," he muttered under his breath. "Could've grabbed one of those meal bars Jennifer bought."

He glanced at the glowing menu board ahead, its cheery photos of breakfast sandwiches and crispy hash browns taunting him. "Alright, this is the last time," he told himself, the words hollow from overuse. He'd said it before—yesterday, last week, last month. Tomorrow would be different, he insisted silently. Tomorrow, he'd skip the line. He'd wake up early, make something healthy, maybe even squeeze in a quick walk before work.

But deep down, he knew better. When the cashier handed him the bag, the familiar smell of grease and salt filled the car, triggering an instant mix of satisfaction and regret. As he pulled back onto the road, he sighed, shaking his head. "Next time," he whispered. "I'll do better next time." But even he wasn't sure he believed it anymore.

By the time Saturday rolled around, Michael craved the energy he once had. Yard work used to be his solace—a time to zone out, sweat a little, and take pride in the freshly mowed lawn. Now the mower seemed heavier, the sun hotter, and his legs more unwilling. He'd drag himself back inside, aching and irritated, collapsing into his favorite recliner. That evening, as the glow of *Saturday Night Live* flickered on the TV, Michael dozed off halfway through the open, the laughter of the audience fading into his dreams. Sleep used to recharge him; now it felt like a fleeting escape.

Jennifer noticed the changes too, although she rarely said much. Intimacy had become a quiet casualty of their shared struggles. Michael couldn't remember the last time they'd been . . . close.

One night, as they were climbing into bed, Jennifer brushed her hand across his shoulder and smiled. "Remember when we used to stay up late, just talking?" she said, her voice warm. Michael chuckled softly, leaning into her touch. "Yeah, and somehow we still made it to work the next morning," he replied, his hand instinctively finding hers. For a moment, the air between them felt lighter, like it used to. But as the quiet stretched, the weight of the day caught up with them. Michael yawned, his body sinking heavily into the mattress. "Long day," he mumbled, closing his eyes. Jennifer didn't respond, and he felt her hand slip away, her warmth retreating as she rolled to her side.

The distance gnawed at him, another thread in the weaving fabric of disappointment and inertia.

Dieting had always seemed like the solution, but every attempt felt like a cruel game. He'd start with enthusiasm, cleaning out the pantry and planning meals. The first few days would go well—he'd skip the drive-through, even lose a pound or two. But then the cravings would come, insistent and loud, and by the end of the week, he'd be back to old habits. The cycle repeated itself, each failure eroding his resolve and reinforcing the idea that change might just be beyond his reach.

Michael had always been good at adapting. When the weight crept on and his energy dipped, he adjusted. He swapped weekend runs for watching sports on TV. He stopped worrying about his old suits gathering dust in the back of the closet. Life was full—work, kids, the endless cycle of responsibilities—and he convinced himself that this was just how things were supposed to be. But underneath his quiet acceptance lay an unspoken longing for something more. He didn't like who he'd become, and although he rarely admitted it, the thought of staying this way terrified him.

The truth was, he thought about the future constantly. He thought about walking Emma down the aisle one day, about swimming with Jacob next summer, about being healthy enough to take trips with Jennifer once the kids were grown. He longed for those things, but they seemed far away, not pressing from where he stood now.

Michael's quiet acceptance wasn't indifference—it was fear. Fear of failing again, of trying and not seeing results, of admitting just how far he'd let himself go. But the distance with Jennifer ignited something in him, a spark he couldn't ignore. She wasn't giving up, and he realized he didn't want to give up either. Sitting alone in the dimly lit living room that night, Michael felt that maybe, just maybe, change was still possible.

Jennifer: The Hopeful Realist

Jennifer stood in the kitchen, scrolling through yet another recipe promising to "Transform your health in 30 days!" Her phone was filled with bookmarks like this—kale smoothies, meal prep plans, beginner yoga videos. She'd tried them all at one point or another. Paleo, keto, intermittent fasting—they always started the same way: with optimism and a fresh set of groceries. And they always ended the same way, too: with frustration, an empty chip bag, and a silent vow to try harder next time.

It wasn't for lack of effort. Jennifer had always been a doer, the kind of woman who approached life with zeal. At 30, she'd opened her own salon, pouring her energy into the business while raising three kids.

Back then, she could juggle it all—work, family, and even an occasional Zumba class with her clients and younger friends. But over the years, her energy had dwindled, and the vibrant, active woman she once was seemed harder to recognize in the mirror.

How did it even get this far? Jennifer thought, staring at the pile of clothes she'd shoved to the back of her closet. Her favorite jeans, the ones that used to hug her just right, now wouldn't budge past her thighs. It hadn't happened overnight—just a few pounds after Emma was born, a few more when Jacob started daycare, and a few more by the time Noah started middle school. At first, she'd chalked it up to the chaos of raising three kids. Stress, hormones, aging—it all sounded reasonable. But reasonable didn't make it easier to accept.

Jennifer sighed, closing the closet door like she could shut out the nagging voice in her head. You've tried. You've really tried. How many diets had it been now? Paleo, keto, Whole30, something with points . . . they all blurred together. For the first week, she'd feel hopeful, even excited, counting calories and sticking to the plan. But then life would happen—a missed meal prep because of an emergency at the salon, or a late-night pizza order after a long day of soccer practice and endless homework battles. The wheels always came off eventually, and with them, her confidence.

Jennifer ran her hands through her hair, the familiar wave of guilt washing over her. Why can't I just stick with it? she thought, her stomach tightening. It wasn't just about the weight; it was about the sense of failure that came with every abandoned plan. She hated the way she avoided mirrors, the way she pulled at her clothes, trying to make them drape just right. And worst of all, the way she felt like she was letting her family down. What kind of example am I setting for Emma? For the boys?

She leaned against the dresser, closing her eyes. It wasn't like she didn't want to be healthy. She wanted to feel good again, to wake up with energy, to see Michael's eyes light up when she walked into a room the way they used to. But every time she tried to make a change, life seemed to get in the way. The diets felt like quicksand, pulling her down with their impossible rules. The fitness trends? A joke. Who had time for spin classes or yoga videos when there was laundry piling up, kids needing rides, and a business to run?

Buried under the frustration was a flicker of determination. I can't keep going like this. I just can't. She wasn't ready to give up—not yet. But she didn't know where to start. All she knew was that something had to change, and soon. For her family. For herself. For the woman she still believed she could be.

Jennifer wasn't just worried about herself, though. She saw it in Michael, too—the way he avoided mirrors, the way he seemed to grow more tired with each passing year. She noticed the little things: his hesitation when climbing stairs, his attempts to skip family outings that required too much movement. She worried about his health, about what their future would look like if they didn't find a way to turn things around. But she also knew pushing him wouldn't work. Michael wasn't the type to respond to pressure; he needed to come to it on his own.

Still, Jennifer wasn't ready to give up. She loved her family fiercely, and she believed in their ability to change, even if it felt impossible some days. Her hope wasn't blind optimism—it was rooted in her belief that small steps could add to something meaningful. If she could just find the right way forward, maybe this time would be different. For herself. For Michael. For the life they still wanted to build together.

Jennifer had always been the multitasker, the one who could juggle a thousand spinning plates without breaking stride. But lately, the balance felt impossible, her days a relentless push and pull between the demands of her career, her family, and her growing desire to take control of her health.

The salon was her pride and joy, a business she'd built from the ground up. She loved helping women feel beautiful, confident, and seen. But it wasn't lost on her how draining that could be, especially when she struggled to feel those things herself. After a long day of standing on her feet, listening to clients vent about everything from work to relationships, she'd come home physically and emotionally spent, with little left to give to her family—or herself.

Making healthy choices was always a challenge. She wanted to eat better, she really did, but who had time to read nutrition labels or Google the hidden sugars in every grocery store aisle? More often than not, dinner was whatever she could grab on the way home after picking the kids up from soccer practice. And forget about meal prepping—her Sunday afternoons were spent managing the chaos of three kids, each with their own unique "emergency."

Parenting, as much as she loved it, was another layer of exhaustion. In the early days, she and Michael had bought into the idea of "cry it out," only to spend countless sleepless nights rocking the babies back to sleep anyway. Now, it was the endless logistics of elementary schoolers and pre-teens—coordinating rides, refereeing arguments, finding sitters they weren't sure they needed for the increasingly rarer date nights with Michael. The sitters were never perfect, and often she'd spend her night worrying about what the kids had eaten or if their actual bedtime had been enforced.

Those date nights weren't what they used to be either. Gone were the spontaneous evenings at John Mayer or Dave Matthews Band concerts, where they'd lose themselves in the music and each other. These days, it was all she could do to stay awake past 10 p.m. And when she did have a moment of quiet, there was always housework looming—laundry that never seemed to end, dishes piled high, and the constant clutter that comes with three kids.

Jennifer knew she needed to prioritize herself, but how? Between her business, her family, and the mountain of daily responsibilities, self-care felt like a luxury she couldn't afford. And yet the longing for something more—for energy, for confidence, for a return to the woman she used to be—never left her. Each day, she carried the hope that change was possible, even as she wrestled with the fear that life might never slow down enough to let her make it happen.

Jennifer sat in the doctor's office, her hands clasped tightly in her lap. She had expected the usual lecture—eat fewer carbs, exercise more, find balance. But her new doctor, Dr. Stevens, was different. He didn't just rattle off generic advice; he took the time to ask questions. How's your energy? What's your biggest challenge when it comes to staying active? By the time he finished, she felt like he actually saw her—not just her weight, but the complexity of her life.

"I think we should explore something a little different," Dr. Stevens said, pulling up a chart on his tablet. "Have you ever considered wearable fitness technology?"

Jennifer blinked. "You mean the watches that track steps?"

"Not just steps," he replied, smiling. "The newer devices monitor your heart rate, sleep patterns, calories burned, even stress levels. Some even give you guided coaching plans based on your data. It's like having a personal trainer on your wrist."

She hesitated. "I don't know. I've tried so many things. What makes this different?"

"It's not about perfection, Jennifer. It's about awareness," he said gently. "You're busy—juggling work, kids, everything. Wearables don't just track what you're doing; they help you see patterns and make small, meaningful changes. Think of it as a tool to guide you, not judge you."

Jennifer felt a flicker of curiosity, mingled with hope. She'd always struggled with the one-size-fits-all approach of diets and fitness plans.

But the idea of something personalized, something that could adapt to her life instead of the other way around? That was new.

Dr. Stevens leaned forward. "You mentioned that Michael's been feeling tired, too. Maybe this could be something you both do together. It's not about competing—it's about accountability. A way to support each other."

She thought about Michael, how they'd drifted into parallel lives instead of walking the same path. Could this be a way to bring them back together? To turn their struggles into a shared journey?

"What if I fail again?" she asked, her voice barely above a whisper.

Dr. Stevens smiled. "Small steps, Jennifer. This isn't about immediate results. It's about finding what works for you. And if something doesn't? You adjust. It's progress, not perfection."

As she left the office, a small pamphlet about wearables tucked in her bag, Jennifer felt a cautious optimism bubbling inside her. For the first time in years, she imagined a way forward—not just for her, but for Michael too. A future where they weren't battling their health alone, but together, guided by something as simple as the numbers on a screen. It wasn't a solution yet, but it felt like the first step toward one. And for Jennifer, that was enough to start.

The Wearable Catalyst

The boxes arrived on a Thursday afternoon, unassuming in their plain packaging but brimming with possibility. Jennifer had insisted on ordering matching wearable fitness devices—a sleek, advanced, watch-type model synced to their mobile phones that tracked everything from steps to sleep, and heart rate to calorie burn. Michael had grumbled at the cost—their insurer did not cover their use—but even he couldn't deny the faint spark of curiosity as they tore open the packaging.

The watches were simple but elegant: slim bands with customizable screens, one in black for Michael and a rose gold version for Jennifer. Jennifer ran her fingers over the smooth surface, marveling at how something so small could claim to transform their lives. Michael, meanwhile, turned the device over in his hands, as if expecting it to whisper the secrets to success.

In addition to the watches, Michael's doctor had suggested the addition of a continuous glucose monitor (CGM) and a connected blood pressure monitor (BPM) to his setup. The CGM was a small, discreet patch that adhered to his arm, syncing with the device to provide real-time blood sugar data. The BPM, a sleek cuff that connected wirelessly, promised to give him better insight into the occasional spikes his doctor had flagged as concerning. Michael eyed the extra devices warily as he unpacked them, feeling a mix of apprehension and determination. Although the watch seemed straightforward enough, these additional tools felt like a step into uncharted territory—more data, more responsibility, and more accountability than he'd ever faced before.

They set up the watches and Michael's other devices together at the kitchen table, surrounded by open instruction booklets and their mobile phones displaying the companion app. The setup was surprisingly intuitive, and within minutes, the devices were synced and glowing faintly on their wrists.

"Welcome to your health journey," the health app displayed after their profiles were completed. Jennifer couldn't help but smile. Michael smirked. "Sounds like a pitch from one of those infomercials," he muttered, but his tone was lighter than usual.

Jennifer clicked through the expanded dashboard on her phone, marveling at the real-time feedback. Her heart rate, activity level, and even a daily step goal were displayed in crisp, colorful graphics. Michael poked at his phone screen more skeptically, his brows furrowing as he navigated the menus. "So, it just . . . magically calculates all this stuff?" he asked. "Not magic," Jennifer said with a laugh. "Technology."

The watches buzzed on their wrists almost simultaneously. "Time to move!" Jennifer read aloud. She stood up eagerly, stretching her arms. Michael glanced at her, then at the screen. "Already bossing me around," he said, rolling his eyes, but he stood too, grudgingly.

Over dinner that night, they found themselves comparing notes. Jennifer was already imagining how she could integrate the device into her daily routine— wearing it at the salon, tracking her sleep, and finally understanding why she always felt so tired. Michael admitted he was curious to see what the sleep tracker would reveal. "If it tells me I snore, I'm returning it," he joked, making her laugh.

As they settled in for the evening, both couldn't help but glance at their wrists every so often. There was a cautious hope in the air, tinged with doubt but underlined by something more profound: the sense that this could be the start of something different. Neither said it out loud, but they both wondered if these small, glowing devices might hold the key to reclaiming not just their health, but their life together.

The first few days with their wearables were a whirlwind of adjustments, marked by moments of discovery and no small amount of frustration. Jennifer dove in with her usual determination, wearing her tracker like a badge of honor and eagerly exploring its features. She experimented with setting step goals, checking her progress throughout the day.

At the salon, she turned the watch into a conversation starter with clients, sharing tidbits about how it tracked her heart rate and calorie burn while she worked. "It even reminds me to stand up and stretch," she told one client, laughing as her watch buzzed during their chat.

For Jennifer, the tracker was an ally—a tool that made her feel in control of her health again. She started small, walking near the salon during quiet moments and taking the stairs anywhere she could. The app quickly became her go-to companion, giving her data she hadn't realized she needed. Her sleep analysis, for instance, revealed she rarely got into deep sleep, something she hadn't fully understood before. "Maybe this is why I'm always tired," she said to Michael one evening, showing him the graph on her phone.

Michael, on the other hand, approached the wearables with caution. The tracker on his wrist was fine; he could handle step goals and the occasional reminder to move. But the CGM and BPM felt intrusive. The CGM's frequent alerts about small blood sugar fluctuations were unnerving at first. "Do I need to do something? Or is it just . . . telling me?" he muttered one afternoon, staring at his phone. Jennifer encouraged him to stick with it, pointing out patterns that were starting to emerge. "Look," she said one evening, "your blood sugar always dips in the late afternoon. Maybe that's why you feel so wiped out around 4 p.m."

Despite his skepticism, Michael couldn't deny that he was intrigued. The apps weren't judging him; they were just showing him the facts. He began to notice small shifts in his awareness—how his morning coffee spiked his blood pressure slightly, or how a longer walk after dinner helped stabilize his blood sugar. The information gave him something he hadn't expected: a sense of control, however small, over his health.

Jennifer's enthusiasm occasionally bordered on overwhelming, but it was hard not to be swept up in her optimism. Noticing the multiuser feature on the health app, she suggested a family step challenge. Michael resisted at first, claiming the kids would beat him easily. But by the third evening, he was out in the yard, pacing alongside Jacob and Emma, counting steps and laughing as they raced ahead.

For both of them, the watches were more than gadgets. They were opening doors—to better understanding, to new habits, and to a future that felt just a little more within reach. For the first time in years, Michael and Jennifer were beginning to feel like they were on the same team again.

The real-time feedback from their watches was a revelation, pulling back the curtain on habits Michael and Jennifer had never fully realized.

For years, they had been guessing—trying diets and routines based on vague advice or fleeting trends. Now, their apps provided answers, shining a light on the intricate patterns of their daily lives.

For Michael, the CGM became an unexpected ally. He had always known his energy levels dipped in the late afternoon, but seeing his blood sugar graph confirmed it: a steady decline starting around 3:30 p.m., often aggravated by his go-to snack of a vending machine candy bar. One evening, he sat with Jennifer, poring over his app. "Look at this," he said, pointing to the graph. "Every time I grab something sugary, it spikes, then crashes even lower." Jennifer nodded, scrolling through her own app. "Maybe swap the candy for some nuts or fruit? It's showing here that protein stabilizes your levels." He tried it the next day, grabbing a handful of almonds instead, and for the first time in years, he didn't feel the usual fog creeping in before dinner.

Jennifer's revelations were more about movement—and how much she wasn't doing. The tracker's step count was a sobering reminder of how her salon shifts, while physically taxing, didn't involve the kind of sustained activity her body needed. "No wonder I feel stuck," she told Michael, pointing to her daily totals. "I'm on my feet all day, but I barely move." She began incorporating short walks during her lunch break, even doing a short walk around the common area before opening the salon. Within a week, her step count had doubled, and she felt a surprising boost of energy by the afternoon.

Even their sleep patterns came under scrutiny. Both Michael and Jennifer had assumed their exhaustion was simply the cost of busy lives. But their watches painted a clearer picture. Michael's sleep tracker revealed a string of restless nights, his sleep often interrupted by brief periods of wakefulness he couldn't remember. "Probably from all that scrolling on your phone before bed," Jennifer teased, but her own information showed she wasn't faring much better. She wasn't getting enough rapid eye movement (REM) sleep, which explained why she still felt groggy even after eight hours in bed. Together, they decided to implement a new rule: no screens an hour before bedtime. It wasn't easy, but within a week, both noticed they were waking up more refreshed.

The devices turned vague aspirations into actionable steps, giving them both a sense of direction they hadn't had before. It wasn't about perfection; it was about progress—small, manageable changes that added up to something bigger. For the first time, they weren't just hoping for change. They could see it, measure it, and, slowly but surely, live it.

A Journey of Wins and Losses

The first signs of progress were small but undeniable. Jennifer noticed it one evening as she cleaned up after dinner. Her usual post-meal fatigue wasn't there, replaced instead by a quiet hum of energy. She'd hit her step goal for the fourth day in a row, something she hadn't thought possible when she first strapped the watch onto her wrist. "Michael," she called, glancing at her wrist, "look. I actually did it." He looked up from the couch, eyebrows raised. "Four days straight? I'm impressed," he said, flashing her a grin.

Michael's victories came in quieter moments. One Saturday morning, as he pulled on an old pair of jeans, he stopped, puzzled. They felt looser. Not dramatically so, but enough that he had to tighten his belt an extra notch. He stood in front of the mirror for a moment, taking in the subtle change in his reflection. "Huh," he muttered to himself, a small spark of pride warming his chest. At breakfast, he shared the news with Jennifer, who responded with a playful poke and a cheerful, "See? You're doing it!"

The watches and the health apps also brought unexpected joys. One evening, as the sun set, Michael and Jennifer decided to take an impromptu walk around the neighborhood. It wasn't a grand gesture—just a few steps around the block—but they walked hand in hand, laughing as they pointed out the patriotic Independence Day decorations on their neighbors' homes. It reminded Jennifer of the early days of their marriage, when evening strolls were a regular occurrence. "We should do this more often," she said softly, and Michael nodded, squeezing her hand.

Even family moments felt different. The kids had taken an interest in the step challenges Jennifer created on the app. One Saturday, they decided to turn it into a family competition, seeing who could rack up the most steps. Michael found himself racing Jacob up and down the backyard, Emma cheering them on from the sidelines. By the time they collapsed on the grass, out of breath and laughing, Michael realized he hadn't felt this lighthearted—or this connected to his kids—in years.

The changes weren't just physical; they were emotional, too. Jennifer felt a renewed sense of control, as if she were finally steering the ship instead of being tossed around by the waves. Michael, for his part, began to see himself differently—not as a man trapped by his past, but as someone capable of change. Each pound lost, each extra lap, each burst of laughter reminded them that transformation wasn't just possible—it was happening. And for the first time in a long time, they felt hopeful, not just for themselves but for each other.

For all their early successes, Michael and Jennifer quickly discovered that progress wasn't a straight line. The enthusiasm of their first weeks began to wane as the reality of their long journey set in. Some days, the setbacks felt heavier than the victories, a reminder of just how deeply their old habits were ingrained.

Michael hit his first wall one evening after a stressful day at work. Deadlines had piled up, and the CEO's curt email about quarterly earnings left him tense and irritable. As he drove home, the glowing sign of the McDonald's drive-through pulled him in like a magnet. He stared at the menu, wrestling with the familiar pull of comfort food. "Just this once," he told himself, ordering his usual combo. By the time he got home, the empty fries carton on the passenger seat was a silent testament to his lapse. When Jennifer noticed the faint smell of grease in the car later, he felt a wave of shame but didn't say anything. That night, he skipped syncing his tracker, unable to face the data.

Jennifer had her own moments of discouragement. One afternoon, as she scrolled through her app, she saw her weight hadn't budged in nearly two weeks. She'd been hitting her step goals, tracking her meals meticulously, and even drinking more water. "What's the point?" she muttered, slamming her phone down on the counter. The frustration spilled into the rest of her day. At the salon, her patience wore thin, and by the time she got home, she was snapping at Michael over whose turn it was to do the dishes.

There was deeper agitation, too—born not from lack of love but from shared frustration. One evening, Jennifer suggested they skip their post-dinner walk, admitting she was too tired. "You're tired?" Michael snapped. "I've been dragging myself through this whole thing, and now you want to quit, too?" His words stung, and she shot back, "Maybe if you'd actually commit instead of sneaking fast food, this wouldn't feel so hard!" The silence that followed was heavy, both retreating to separate corners of the house to cool off.

And then there were the moments when the weight of their past felt like a shadow they couldn't outrun. Michael found himself staring at old family photos, where his slimmer, younger self looked carefree in a way that still felt far away. Jennifer caught herself comparing her reflection to photos of friends on social media, feeling a pang of envy she hated to admit.

Despite the setbacks, neither gave up. But these moments served as reminders that transformation wasn't just physical—it was emotional and deeply tied to the stories they told themselves about who they were and what they could achieve. Some days, those stories felt like the hardest weight of all to lift.

The turning point came one evening after another difficult day. Michael and Jennifer sat on the couch, the remnants of dinner still on the table, their watches buzzing with unmet step goals. The silence between them was heavy, but it wasn't the kind of silence that pushed them apart—it was the kind that pulled them closer, an acknowledgment of the shared struggle.

Jennifer broke the quiet first, sighing deeply as she tucked her legs under her. "This is harder than I thought it'd be," she admitted, her voice soft. Michael nodded, running a hand through his hair. "Yeah. Feels like every time I make a little progress, something drags me back." He looked at her then, his usual defenses lowered. "I . . . I can't do this alone."

Jennifer reached for his hand, her fingers curling around his. "You don't have to," she said simply. The words hung in the air, heavier with meaning than she'd realized. "We're in this together. We can't let the bad days undo the good ones."

From that moment, they started leaning on each other in ways they hadn't before. When Michael felt the pull of fast food, he'd text Jennifer instead, a simple "bad day" message that was met with encouragement instead of judgment. Jennifer, in turn, began sharing her frustrations more openly, admitting when the scale didn't budge or when she felt like giving up. Together, they'd talk it through, finding solutions instead of stewing in their doubts.

Their post-dinner walks became a ritual again, but this time with a new purpose. Instead of focusing solely on step counts, they used the time to talk—about their kids, their plans for the weekend, or dreams they'd shelved for too long. One night, as they strolled under the streetlights, Jennifer squeezed Michael's hand. "You know," she said, smiling, "I think we're getting better at this. At . . . us."

Michael laughed, a deep, genuine sound that Jennifer hadn't heard in months. "Who knew a couple of gadgets could do all this?" he said, gesturing to the tracker on his wrist. But he knew it wasn't just the wearables—it was them, choosing each other over and over, even on the hard days.

They found strength in the small victories: high-fiving after a family step challenge, cooking a healthy meal together, or simply collapsing into bed at night, proud that the day's stresses hadn't derailed them. Their shared commitment wasn't just about weight loss anymore; it was about rediscovering the partnership they'd built years ago, one step, one choice, one moment at a time.

Partnership Renewed

Months into their journey, Michael and Jennifer found themselves in a place they hadn't been in years—connected, not just as parents or partners but as a team. The changes they had pursued together weren't just about their health; they had quietly redefined their marriage, strengthening a bond that had felt stretched thin under the weight of years of routine and responsibility.

It showed in the small things first. Their nightly walks, once a chore to rack up steps, had become a cherished part of their day. Hand in hand, they walked through their neighborhood, talking freely, sometimes laughing, sometimes sharing things they hadn't talked about in years. Michael opened up about his struggles with confidence and agitation, and Jennifer spoke honestly about the toll of juggling her career and family. In those quiet moments, they found a deeper understanding of each other—a closeness they hadn't known they'd lost.

Their teamwork extended beyond the walks. Grocery shopping became a joint effort, a surprisingly fun venture as they swapped out processed foods for healthier options, challenging each other to find new recipes to try. They even rediscovered the joy of cooking dinner together, dancing playfully in the kitchen to their favorite songs. "You still can't dance," Jennifer teased one night as Michael spun her around, his laughter filling the room.

Even their children noticed the change. Noah joked about how lame it was to see his parents walking together every night, and Emma beamed with pride when Michael joined her for an impromptu dance video on TV, his energy levels higher than they'd been in years.

The most profound change, though, was in the way they looked at each other. Gone were the glances tinged with frustration or resignation. In their place was something softer, warmer—a mutual respect for what they had accomplished together and an appreciation for the love that had carried them through the hardest moments.

One evening, as they sat on the porch, watching the sunset, Michael reached for Jennifer's hand. "We've come a long way," he said quietly. Jennifer smiled, leaning her head on his shoulder. "We really have." In that moment, it wasn't just their health that felt transformed—it was their life together, renewed and stronger than ever.

Several weeks later, the evening air was cool but not chilly, carrying the soft hum of chatter and the occasional laughter from families scattered across the park. Michael and Jennifer sat side by side on a picnic blanket, surrounded by the flickering glow of string lights that lined the park's pathways. The small stage at the center of the lawn held a local band playing a mellow acoustic set, their music floating through the air like a shared heartbeat.

For the first time in years, Michael felt at ease, not just physically but emotionally. He stretched his legs out in front of him, the simple act of sitting on the ground feeling strangely freeing. A few months ago, he might have avoided this outing altogether, dreading the discomfort of folding his larger frame onto a picnic blanket or struggling to rise gracefully afterward. Tonight, though, he didn't think twice about it. He leaned back on his hands, the tension that once lived permanently in his shoulders now a distant memory.

Jennifer smiled at him, her cheeks faintly rosy from the crisp air. She looked radiant, more relaxed than he'd seen her in years. "This was a good idea," she said softly, taking in the evening. "We deserve this."

Michael nodded, watching her as the music swelled. She looked so much like the woman he'd fallen in love with all those years ago, not because of any physical transformation but because of the lightness in her expression—the joy that had returned to her eyes. He reached over, threading his fingers through hers.

Around them, the world seemed to slow. Kids ran in the distance, their laughter blending with the music, while couples swayed together under the trees. But Michael and Jennifer stayed rooted in their quiet moment, soaking in the simplicity of being together, free from the weight of their past struggles.

As the band played a familiar song—one they used to dance to during their dating days—Jennifer's eyes lit up. "Remember this?" she asked, squeezing his hand. Michael chuckled. "Are you kidding?" He tugged her to her feet, and they swayed together, their laughter mingling with the music.

For the first time in years, they felt truly present—not defined by who they had been or what they had endured, but by who they were now. Stronger. Closer. Free.

A Hopeful Future

Michael and Jennifer sat together on the porch swing, the sun dipping low on the horizon, casting the world in a warm, golden light. The air was quiet except for the soft creak of the swing and the occasional chirp of crickets. Michael glanced down at the watch on his wrist, its faint glow a reminder of how much their lives had changed.

"It's strange," he said, breaking the silence. "I used to think this was all impossible. That it was too hard to make a difference."

Jennifer smiled, her hand resting lightly on his arm. "I felt the same. Like every attempt was just another chance to fail." She turned to look at him, her eyes warm. "But this was different. These simple watches," she laughed, "didn't just track things. They showed us the truth about where we were and helped us see where we wanted to be."

Michael nodded, his gaze drifting to the yard where their kids had once played. "It wasn't just about the weight or the numbers. It was about seeing the patterns, the habits we didn't know we had. And doing it together," he added, his voice soft. "That made all the difference."

Jennifer leaned her head on his shoulder, a contented sigh escaping her lips. "We've come so far," she said quietly. "Not just in how we feel, but how we are with each other. I never even thought about getting back to this."

Michael smiled, squeezing her hand. "Neither did I. But I'm glad we did. It feels like . . . like we're kids again."

As the sun set, they sat in peaceful silence, grateful for the journey that had brought them not just back to health, but back to each other.

Wearable Devices

Chronic conditions have become a critical healthcare priority, driving escalating costs, reduced quality of life, and preventable deaths.[1] Addressing these challenges demands fundamental change, shifting focus from acute care to prevention, early intervention, and long-term management. Innovations in precision medicine, patient education, and access to care are essential to curb the rising burden of chronic diseases. Policymakers, healthcare providers, and communities must collaborate to create sustainable, patient-centered solutions, ensuring that chronic condition management becomes a cornerstone of modern healthcare and a pathway to healthier populations.

In this chapter, we explore how emerging solutions, such as wearable technology and precision medicine, are revolutionizing chronic disease management. These innovations enable personalized, data-driven approaches to prevention and treatment, empowering patients and providers with real-time insights and tailored interventions. Wearables offer continuous monitoring of vital signs and lifestyle patterns, and precision medicine leverages genetic and environmental data to optimize care. Together, these advancements represent a transformative shift in addressing the challenges of chronic conditions, promising more effective, accessible, and proactive healthcare solutions.

Chronic Conditions: A Growing Concern

Chronic diseases like obesity are long-lasting health conditions that persist for extended periods, often for a lifetime, and typically develop gradually. They include ailments, as in Michael and Jennifer's case, like diabetes, heart disease, and even arthritis. In contrast, acute conditions have a sudden onset, are usually short in duration, and often resolve with appropriate treatment. Examples of acute conditions include the common cold, influenza, and a broken bone. Whereas acute conditions may require immediate attention, chronic diseases necessitate ongoing management to control symptoms and prevent progression.

Chronic conditions are increasingly prevalent, affecting nearly half of adults worldwide, with numbers expected to rise due to aging populations and lifestyle factors.[2] Diseases like diabetes, cardiovascular disorders, and chronic respiratory illnesses are now leading causes of death globally (see Figure 2-1). These conditions account for the majority of healthcare spending, straining systems with long-term care needs, frequent hospitalizations, and high medication costs. The burden is aggravated by differences in access to care, making prevention, early intervention, and innovative management strategies critical to addressing the growing crisis.

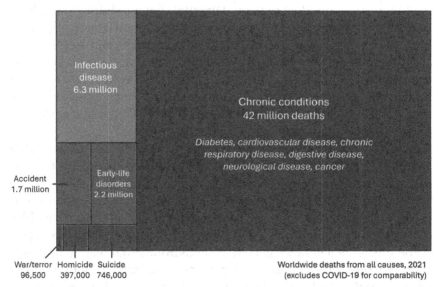

Figure 2-1: Prevalence and impact of chronic conditions

Source: Adapted from [3]

Chronic diseases also pose a central healthcare challenge due to their complex nature, long duration, and significant impact on individuals and systems alike. Unlike acute illnesses, chronic conditions require continuous management, integrating medical, behavioral, and lifestyle interventions. They are primary drivers of healthcare costs, workforce productivity loss, and patient morbidity. Addressing chronic diseases necessitates a shift from reactive to proactive care, emphasizing prevention, early detection, and innovative solutions like precision medicine. Their prevalence underscores the urgent need for comprehensive changes to mitigate their toll on global health and economies.

Lifestyle factors play a significant role in the development and progression of chronic conditions. A poor diet, often characterized by excessive consumption of processed foods, added sugars, and unhealthy fats, contributes to obesity, diabetes, and cardiovascular diseases. Deficiencies in essential nutrients further impair metabolic functions, increasing health risks. Physical inactivity compounds these problems by slowing the metabolism, weakening cardiovascular and musculoskeletal health, and increasing susceptibility to weight gain and insulin resistance. Chronic stress is another critical driver, triggering hormonal imbalances and inflammatory responses that elevate the risk of hypertension, heart disease, and mental health disorders. Persistent stress also encourages unhealthy coping behaviors, such as overeating, smoking, or alcohol consumption, creating a vicious cycle that worsens overall health. Addressing these factors requires a holistic approach, incorporating education, access to healthier food options, encouragement of physical activity, and stress management techniques. Individual behavior changes combined with support from family and friends are key to reducing the burden of chronic diseases rooted in lifestyle choices.

Environmental factors significantly influence the risk of chronic diseases as well, with pollution and urban living playing prominent roles. Air pollution, including particulate matter from vehicles, industrial emissions, and natural sources, is strongly linked to respiratory conditions like asthma and chronic obstructive pulmonary disease (COPD), as well as cardiovascular diseases. Prolonged exposure to polluted air increases systemic inflammation and oxidative stress, exacerbating existing health conditions and contributing to new ones. Urban living often compounds these risks through limited green spaces, noise pollution, and heightened exposure to harmful environmental toxins. Crowded urban settings may restrict opportunities for physical activity, promoting sedentary lifestyles that increase the likelihood of obesity, diabetes, and heart disease. Additionally, urban heat islands—areas with elevated temperatures due to dense infrastructure—can worsen health outcomes, particularly for vulnerable populations. Addressing these environmental challenges requires coordinated community efforts, including reducing emissions, promoting sustainable urban planning, and increasing access to clean air and green spaces. Such interventions are crucial to mitigating the environmental drivers of chronic disease risk but are often hard-won and challenging to implement.

> **NOTE** Genetics is not destiny. Things we control play integral roles in how our genes present in our daily lives.

Genetic predisposition also plays a critical role in the development of chronic conditions, influencing an individual's susceptibility to diseases like diabetes, cardiovascular disorders, and certain cancers. Specific genetic variations can alter how the body processes fats, sugars, or inflammation, creating an inherent vulnerability. However, genetics is not destiny; the interplay with lifestyle and environmental factors often determines whether these predispositions manifest as disease. For instance, a person with a family history of diabetes may carry genes associated with insulin resistance, but poor dietary habits, physical inactivity, or chronic stress can trigger or accelerate disease onset. Similarly, individuals genetically predisposed to hypertension might see this risk heightened by environmental factors like air pollution or high sodium diets.

This interplay underscores the importance of personalized medicine, which leverages genetic information to tailor interventions. By addressing modifiable factors such as diet, exercise, and exposure to harmful environments, individuals can mitigate genetic risks, turning potential vulnerabilities into opportunities for prevention and better health outcomes.

Emotional and Psychological Impacts

Chronic diseases often impose a significant emotional and mental toll on individuals, with conditions like depression, anxiety, and social isolation being common. The prolonged nature of these illnesses, coupled with physical discomfort and limitations, can lead to feelings of hopelessness and frustration. Depression frequently arises from the ongoing burden of managing symptoms, navigating healthcare systems, and coping with lifestyle changes. Anxiety is another common response, fueled by concerns over disease progression, financial strain, and the uncertainty of treatment outcomes.

Social isolation can worsen these emotional challenges, as individuals may withdraw due to stigma, reduced mobility, or the inability to participate in previously enjoyed activities. This isolation further diminishes quality of life and complicates disease management. Addressing the psychosocial dimensions of chronic diseases requires comprehensive care models that integrate mental health support, community resources, and peer networks. These interventions help patients build resilience, reduce emotional suffering, and enhance overall well-being.

Societal stigma surrounding chronic conditions, such as obesity, mental health disorders, and HIV/AIDS, can significantly affect patients' well-being and treatment outcomes. This stigma often stems from misconceptions, moral judgments, or a lack of understanding about the complex causes of these diseases. For instance, individuals with obesity may be unfairly blamed for their condition, despite underlying genetic, metabolic, or environmental factors. Similarly, those with mental health disorders often face discrimination that discourages them from seeking help. The impact of stigma is profound, leading to feelings of shame, low self-esteem, and isolation. Patients may delay or avoid medical care due to fear of judgment, aggravating their conditions. Stigma also influences public policies, potentially limiting funding for research or access to care for stigmatized diseases.

Reducing stigma requires widespread education, public awareness campaigns, and empathetic healthcare practices that emphasize understanding and support over judgment, enabling patients to seek care without fear.

Similarly, strong support systems are vital in managing and treating chronic conditions, as they provide the emotional, practical, and medical assistance needed for patients to navigate their health journeys. Family members often play a central role by offering daily care, encouragement, and accountability, helping patients adhere to treatment plans and make necessary lifestyle changes. Communities also contribute by fostering social connections that reduce isolation and providing access to support groups, educational resources, and local services.

Healthcare providers are essential members of this network, delivering personalized care, monitoring progress, and coordinating interdisciplinary interventions. They also empower patients with knowledge and tools to actively participate in their treatment. Together, these support systems enhance resilience, improve treatment adherence, and promote a sense of empowerment. Healthcare systems, however, have traditionally emphasized acute care, prioritizing the treatment of immediate, short-term conditions like infections or injuries over the ongoing management of chronic diseases. This acute-care model, although effective for emergencies, often falls short in addressing the complex, long-term needs of chronic conditions such as diabetes, heart disease, and arthritis.

One key issue is the episodic nature of acute care, which focuses on isolated events rather than continuous, holistic patient engagement. Chronic diseases require sustained attention, lifestyle interventions, and preventive strategies that are often overlooked in systems designed for acute crises. Additionally, fee-for-service payment models incentivize procedures and hospital visits rather than proactive, integrated care, perpetuating the cycle of reactive treatment.

The neglect of chronic disease management contributes to higher healthcare costs, poorer patient outcomes, and preventable complications. Patients with chronic conditions often experience fragmented care, with multiple providers and inconsistent follow-ups, leading to gaps in treatment and a lack of coordinated strategies.

To address this imbalance, healthcare systems must shift toward value-based care models emphasizing prevention, early intervention, and multidisciplinary approaches. Initiatives implementing value-based care are advancing, to be sure, but integrating chronic disease management into primary care and leveraging technology for monitoring and patient engagement can accelerate closing of this gap, ensuring comprehensive, patient-centered care.

NOTE Gaps in our knowledge significantly hinder our ability to address chronic conditions.

Gaps in patient education and access to resources also significantly hinder effective management of chronic conditions. Many patients lack a clear understanding of their condition, treatment options, and the lifestyle changes required for long-term control. This knowledge deficit often results from time-constrained medical appointments, where providers focus on immediate concerns rather than comprehensive education. Without proper guidance, patients may struggle to adhere to treatment plans, leading to worsening symptoms and preventable complications.

Sociological Factors

Access to essential resources is another critical barrier. Socioeconomic differences often limit access to medications, nutritious food, exercise facilities, and technology like wearable devices or remote monitoring tools. Rural and underserved areas face additional challenges, including fewer healthcare facilities, limited specialist availability, and poor health literacy infrastructure. These gaps disproportionately affect vulnerable populations, intensifying health differences and perpetuating cycles of poor chronic disease outcomes.

Addressing these issues requires multifaceted interventions. Providers must prioritize patient-centered communication, using accessible language and culturally sensitive materials to empower patients. Community programs, telemedicine, and public health initiatives can expand access to necessary resources, ensuring reasonable support for self-management. Strengthening these educational and resource frameworks is essential for reducing chronic disease burdens and improving overall health outcomes.

Healthcare differences significantly affect underserved populations, too, posing disproportionate challenges in managing chronic diseases. Socioeconomic factors such as income, education, and employment status limit access to quality care, nutritious food, safe exercise environments, and necessary medications. Individuals in rural areas or low-income urban neighborhoods often lack nearby healthcare facilities, making consistent management of chronic conditions like diabetes, hypertension, and heart disease difficult.

Insurance coverage differences further compound these challenges. Underserved populations are more likely to be uninsured or underinsured, resulting in delayed care, skipped medications, or reliance on emergency services for preventable complications. Cultural and language barriers can also hinder communication between patients and providers, leading to misunderstandings about treatment plans or missed opportunities for preventive care.

These differences contribute to worse health outcomes, higher rates of complications, and increased mortality in vulnerable groups. Addressing this requires targeted interventions, including expanding access to affordable care, improving health literacy, and promoting culturally competent practices. Community-based programs, telehealth, and other services focused on availability are essential for leveling the playing field and ensuring that underserved populations can effectively manage chronic diseases and achieve better health outcomes.

A shift toward proactive and preventative care is essential to address the growing burden of chronic conditions. By prioritizing early detection, lifestyle interventions, and regular monitoring, healthcare systems can reduce the prevalence and severity of these diseases. Preventative care strategies, such as screenings, vaccinations, and personalized health plans, empower individuals to take control of their health before complications arise. This approach not only improves patient outcomes but also decreases healthcare costs by minimizing the need for costly acute treatments. Investing in prevention is a sustainable path to healthier populations and more efficient care systems.

Wearables and Precision Medicine: Tailoring Health Solutions

At their core, wearable devices are compact, technology-enabled tools designed to monitor and track various health metrics in real time. Common examples include fitness trackers, smartwatches, and medical-grade sensors; rings, bands, and even clothing are becoming increasingly available in the market. These devices measure parameters such as heart rate, activity levels, sleep patterns, and blood oxygen levels, providing users and healthcare providers with continuous data. Advanced wearables can even detect irregular heart rhythms, monitor glucose levels, and track stress indicators. By offering personalized insights, wearables empower individuals to make informed lifestyle choices and enable proactive health management, making them a cornerstone of modern, data-driven healthcare.

NOTE   Wearable devices link real-time, user-specific data to the targeted solutions of precision medicine.

Like wearables, precision medicine is a transformative healthcare approach that tailors treatments and preventive strategies to the unique characteristics of each individual. Precision medicine aligns seamlessly with advancements like wearables, offering a data-driven, patient-centered framework that empowers individuals and healthcare providers to manage health conditions more effectively and proactively.

Wearables complement precision medicine by delivering continuous, individualized health data that enriches personalized care strategies. These devices monitor vital signs, physical activity, and other metrics in real time, offering insights into a patient's daily habits and physiological responses. By integrating this data with genetic and environmental information, precision medicine creates a comprehensive health profile for targeted interventions. Wearables also enhance patient engagement, empowering individuals to actively manage their health. Together, these technologies enable more accurate diagnoses, dynamic treatment plans, and proactive prevention of chronic conditions. This comprehensive approach empowers users and healthcare providers to make informed, personalized decisions, optimizing prevention and treatment plans in real time while supporting broader precision medicine initiatives.

For example, the data collected by wearables is also instrumental in tailoring individual health solutions, enabling dynamic and precise care. Continuous monitoring of metrics like heart rate, blood pressure, and glucose levels provides real-time insights into a patient's health, allowing healthcare providers to adjust medications for optimal effectiveness and minimal side effects. For instance, wearables can alert clinicians to irregular heart rhythms, prompting timely changes in treatment or interventions.

Lifestyle recommendations are similarly customized using wearable data. Activity and sleep patterns inform tailored fitness plans, and stress metrics guide mindfulness or relaxation techniques. Wearables also track adherence to prescribed regimens, helping users and providers identify gaps and refine strategies for better outcomes.

By integrating wearable data with genetic, environmental, and clinical information, precision medicine leverages these insights to deliver proactive, personalized solutions. This approach not only enhances treatment efficacy but also empowers individuals to make informed decisions, fostering long-term health improvements.

Wearable devices are transforming chronic disease management and early detection through real-world applications. For instance, continuous glucose monitors (CGMs) are revolutionizing diabetes care by providing real-time blood sugar data, enabling patients and providers to fine-tune insulin therapy and dietary plans. Similarly, smartwatches with ECG capabilities, like the Apple Watch, have been instrumental in detecting atrial fibrillation, a leading cause of stroke, allowing for early intervention and treatment.

In hypertension management, wearables that monitor blood pressure over time help track trends and adjust medications accordingly. Devices like the Oura Ring, which tracks sleep and recovery metrics, aid individuals with sleep apnea or chronic fatigue by identifying irregular sleep patterns and suggesting interventions.

Additionally, wearables play a growing role in preventive care, with stress monitors helping users manage anxiety levels and prevent burnout. These examples demonstrate the potential of wearable technology to improve outcomes and empower proactive health management.

NOTE   Wearable devices bridge the knowledge gap between patients, their circumstances, and their health.

Wearables empower patients by delivering real-time feedback on key health metrics, fostering greater awareness and engagement in their well-being. Devices tracking physical activity, heart rate, and sleep patterns provide immediate insights, helping users understand how daily behaviors impact their health. This continuous feedback creates a sense of accountability, motivating individuals to adopt healthier habits, such as increasing exercise, improving diet, or prioritizing sleep. Real-time alerts and personalized recommendations further enhance behavior change. For example, wearables can prompt users to stand up after prolonged sitting, encourage hydration, or signal early warning signs of stress or abnormal heart rhythms. This immediacy empowers patients to make informed decisions and take corrective actions before problems escalate.

By transforming passive health monitoring into an interactive experience, wearables strengthen patient commitment to their goals. This proactive engagement not only improves adherence to treatment plans but also fosters lasting lifestyle changes, essential for managing chronic conditions and optimizing health outcomes. Indeed, seeing immediate results from wearable devices has profound psychological effects, fostering accountability and promoting habit formation. Real-time feedback on metrics such as steps taken, calories burned, and heart rate during exercise provides users with a tangible sense of progress. This instant gratification reinforces positive behaviors, making individuals more likely to repeat and sustain them. Accountability is a critical key benefit, not to be underestimated. Wearables often include goal-setting features and daily reminders, helping users stay consistent. Knowing their activity is being tracked motivates individuals to adhere to their health plans, transforming abstract objectives into actionable daily habits.

Additionally, seeing small, incremental improvements, like better sleep scores or a lower resting heart rate, creates a rewarding feedback loop. This reinforces the belief that their efforts are paying off, boosting self-efficacy and commitment. Over time, these repeated behaviors solidify into lasting habits, empowering users to maintain healthier lifestyles and achieve long-term health goals.

Wearable Device Limitations

Wearable devices, although transformative, have limitations that can impact their effectiveness. Variability of accuracy is a significant concern, as not all wearables provide precise measurements for metrics like heart rate, calorie expenditure, or sleep quality. Differences in sensor technology, skin tone, and device placement can lead to inconsistencies, potentially causing users to misinterpret their health data.

Technical issues, such as connectivity problems, short battery life, and software glitches, can also hinder usability. For instance, interrupted data syncing with smartphones or health platforms may disrupt real-time monitoring and insights. Additionally, wearables may lack standardization, making it challenging to integrate their data with electronic health records or coordinate with healthcare providers.

These limitations highlight the need for ongoing technological improvements and rigorous validation of wearable devices. Wearable data should be viewed as a supportive tool rather than definitive diagnostic evidence, complementing professional medical advice for a holistic approach to health management.

Wearable devices also can raise significant concerns regarding the collection, storage, and use of health data. As these devices continuously monitor sensitive metrics like heart rate, activity levels, and sleep patterns, questions about data ownership and control become critical. Users often lack clarity about who owns their data—whether it is the individual, the device manufacturer, or a third-party platform that processes the information.

The potential for misuse further complicates these concerns. Health data could be exploited for targeted advertising, shared with employers or insurers without consent, or even exposed in data breaches, risking personal and financial harm. The aggregation of such data also raises ethical questions about its use in research or commercial ventures without user awareness.

To address these challenges, robust regulations and transparent policies are essential. Users should have full control over their data, with options to consent to specific uses. Ensuring strong encryption and secure data practices is crucial to protecting privacy and maintaining trust in wearable technology.

Access differences present a significant barrier to the integration of wearables in healthcare as well. High costs often limit these devices to wealthier individuals, leaving low-income populations unable to benefit from their transformative potential. Similarly, rural and underserved communities may lack the technological infrastructure, such as reliable internet access, necessary for wearable functionality and data integration with healthcare systems.

Bridging the Gaps

These differences aggravate existing healthcare gaps, as populations already facing limited access to care are excluded from advancements that could improve chronic disease management and preventive care.

Reasonable integration of wearables requires addressing affordability through incentives, insurance coverage, or lower-cost alternatives. Public health initiatives must consider availability for underserved groups, ensuring that wearables are accessible across socioeconomic and geographic lines. Additionally, education and support programs should accompany devices to enhance usability and engagement. By making wearables inclusive, healthcare systems can close gaps and promote broader health achievement.

Wearables play a critical role in bridging gaps in traditional healthcare by providing personalized and preventive care, addressing limitations in one-size-fits-all approaches. These devices continuously monitor individual health metrics, offering real-time data that enables tailored interventions. For example, wearables can alert users to early warning signs of chronic conditions, such as irregular heart rhythms or elevated stress levels, prompting timely preventive actions.

In underserved or rural areas, wearables expand access to care by facilitating telehealth consultations and enabling remote monitoring, reducing the need for frequent in-person visits. They also empower patients to take an active role in their health, fostering engagement and adherence to personalized treatment plans. By integrating wearable data with precision medicine strategies, healthcare systems can shift focus from reactive treatment to proactive management, ultimately improving outcomes and reducing differences in care accessibility and quality.

Wearables occupy a unique position in integrating patient-generated data into broader healthcare ecosystems, creating a bridge between individual health behaviors and institutional medical care. These devices collect continuous, real-time data, providing a detailed picture of daily health trends. When integrated with electronic health records (EHRs), this data offers healthcare providers valuable insights into patients' lifestyles and conditions outside of clinical settings. By enabling remote monitoring, wearables enhance chronic disease management and support early intervention. They also complement telehealth services, allowing clinicians to make informed decisions based on up-to-date, patient-specific information. Furthermore, aggregated wearable data can inform population health strategies, aiding research and the development of tailored healthcare policies. Through their seamless integration with digital health platforms, wearables empower patients and providers alike, advancing a more connected, data-driven approach to personalized healthcare.

Wearables are pivotal enablers of precision medicine, unlocking its potential to revolutionize healthcare by delivering real-time, personalized insights into individual health. By continuously tracking key metrics, wearables provide the granular data needed to tailor interventions, optimize treatments, and prevent disease progression. Their ability to integrate patient-generated data with genetic, environmental, and clinical information creates a comprehensive, individualized health profile, allowing precision medicine to thrive.

Beyond individual care, wearables support broader healthcare advancements, such as remote monitoring and population health research, making precision medicine accessible to diverse populations. They empower patients to actively engage in their health management while equipping providers with actionable insights to make informed decisions.

As technology advances, wearables will further bridge the gap between innovative medical approaches and everyday care, transforming healthcare into a proactive, personalized system. Their role underscores the promise of precision medicine to improve outcomes, enhance health achievement, and redefine modern healthcare.

A HISTORICAL PERSPECTIVE ON WEIGHT MANAGEMENT

Humans have wrestled with weight management for centuries, since long before the days of calorie-tracking apps and boutique fitness studios. From ancient feasts followed by Spartan fasts to Renaissance ideals of "plump prosperity," our relationship with weight has been anything but simple. Across eras, managing those extra pounds often involved bizarre remedies, peculiar diets, and a universal truth: the struggle between indulgence and restraint is as old as time itself.

Take 19th-century poet Lord Byron, who famously dabbled in extreme dieting, surviving on vinegar and hard, dry biscuits to maintain his "romantic" pallor. His obsession wasn't unique—history is full of such antics, proving humanity's long-standing battle with body image. Whether wrapping in cabbage leaves, chewing each bite 100 times, or declaring fat a "moral failing," our ancestors remind us that the quest for the perfect figure has always been a mix of comedy and tragedy.

In this sidebar, we take a playful yet insightful look at humanity's creative—and often peculiar—attempts to tackle obesity throughout history. From medieval potions promising to "melt away fat" to Victorian-era corsets squeezing the pounds into submission, our ancestors left no stone unturned in their quest for weight loss. As we explore these wacky remedies, prepare to wonder—and cringe—at the enduring ingenuity of our eternal battle with the bulge.

Continues

(Continued)

Ancient Approaches: Philosophers and Potions

The ancient Greeks were among the first to link weight management with health, emphasizing the importance of balance in diet and exercise. Philosophers like Hippocrates, often called the father of medicine, believed that moderation was the cornerstone of good health. He advised against overindulgence and recommended regular physical activity to maintain a harmonious body.

However, not all ancient approaches were so rational. Some societies relied on potions, rituals, and even divine intervention to shed unwanted pounds. Egyptian texts mention weight-loss concoctions made of laxative herbs, and other cultures sought magical cures, reflecting both the ingenuity and desperation of early weight management efforts. These ancient attitudes laid the groundwork for future theories, blending science, superstition, and the ever-present human desire for a quick fix to an age-old struggle.

Ancient weight-loss remedies were as inventive as they were bizarre. The Romans, for instance, embraced sweat baths, or *thermae*, as shown in Figure 2-2, believing that excessive sweating could purge the body of impurities—and, perhaps, extra pounds. Patrons would stew in steamy chambers, hoping to sweat their way to a leaner physique while indulging in lively debates or gossip.

Figure 2-2: Sweating away the pounds in Ancient Rome
Source: Generated with AI using DALL·E - OpenAI

Herbal concoctions were another popular option across ancient civilizations. The Egyptians brewed potions from laxative plants like castor oil, aiming to "cleanse" the body. Meanwhile, in traditional Chinese medicine, teas made from hawthorn berries and other herbs promised to reduce fat by improving digestion.

Although based on limited understanding, these remedies highlight humanity's relentless creativity in pursuing weight loss. Whether sweating in marble baths or sipping dubious brews, ancient people were clearly willing to go to great (and sweaty) lengths to tip the scales in their favor.

The history of weight loss wouldn't be complete without some truly outlandish practices, like the "miracle" ingredients that promised dramatic results. Take vinegar: Lord Byron swore by it in the 19th century, but the practice dates back to ancient times, with hopeful dieters chugging it to cleanse their bodies—although mostly ending up with stomach aches and sour dispositions.

Then there's camel milk (see Figure 2-3), revered by some nomadic cultures as a cure-all, including for weight management. It was believed to balance the humors and slim the body, although the results were likely as mythical as the camel diet itself.

Figure 2-3: A camel milk a day keeps the doctor away.
Source: Generated with AI using DALL·E - OpenAI

These practices prove one thing: from ancient apothecaries to modern influencers, people have always fallen for the latest miracle cure. If nothing else, history shows that desperation has a surprising way of turning everyday pantry items into weight-loss holy grails!

Continues

(Continued)

The Middle Ages: When Fat Was Fashionable

In the Middle Ages, societal attitudes toward body image took a dramatic turn, with excess weight becoming a symbol of wealth, power, and prestige. Unlike earlier periods that prized moderation, medieval society often celebrated a fuller figure as evidence of prosperity. Only the wealthy could afford feasts of meat, bread, and ale, making a robust physique a status symbol. Paintings and sculptures of the era frequently depicted plump nobles and rotund monarchs, reinforcing the association between girth and grandeur. In this period, being fat wasn't just acceptable—it was fashionable!

Although medieval society often celebrated excess weight as a sign of wealth, early medical theories like the four humors painted a less flattering picture. Physicians believed obesity stemmed from an imbalance, particularly an excess of "phlegm" or "blood," which disrupted the body's natural harmony. Treatments ranged from bloodletting, as shown in Figure 2-4, to purging, aiming to restore equilibrium rather than address diet or lifestyle. This contrast highlights the era's mixed attitudes: societal norms admired plumpness, but medical practitioners viewed it as a physiological problem requiring correction—often with dubious remedies.

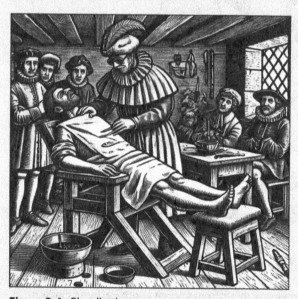

Figure 2-4: Bloodletting
Source: Generated with AI using DALL·E - OpenAI

Medieval treatments for obesity often bordered on the bizarre. Bloodletting was a go-to remedy, with physicians cheerfully draining patients in the hope of lightening both their humors and their waistlines. Meanwhile, odd fasting rituals abounded, including subsisting solely on holy water or bread blessed by monks, with the belief that divine intervention would melt away the pounds. Some even swore by wearing charms or relics thought to curb appetite. These quirky approaches reveal a time when science and superstition collided in the eternal quest to shed weight.

The Victorian Era: Corsets, Tonic Waters, and Tapeworms

The Victorian era brought with it an obsession with achieving the coveted "wasp waist," leading to the widespread use of corsets (see Figure 2-5). These tightly laced garments physically reshaped the body, cinching the waist to sometimes dangerously small proportions. Women endured restricted breathing and even displaced organs to conform to the ideal silhouette, often with a fainting couch nearby for relief.

Figure 2-5: Corseted woman
Source: Generated with AI using DALL·E - OpenAI

The Victorian era saw a boom in patent medicines and tonics, marketed as miracle cures for weight loss. These concoctions, often sold under enticing names like "Dr. Smith's Fat Reducer" or "Flesh Dissolving Elixir," promised effortless slimming. The reality, however, was far less miraculous. Many of these products contained dubious—and downright dangerous—ingredients such as arsenic, strychnine, and cocaine. Although they might suppress appetite or boost energy (at least temporarily), they also carried severe side effects, from poisoning to addiction.

Continues

(Continued)

Manufacturers capitalized on the lack of regulation and society's growing obsession with weight loss, often masking their formulas behind proprietary "secrets." Consumers, desperate for quick fixes, rarely questioned their efficacy or safety. These tonics highlight an era of unchecked marketing and a willingness to sacrifice health for the elusive promise of a slimmer figure—a cautionary tale still relevant today.

Of all the Victorian weight-loss fads, the tapeworm diet takes the cake— and promptly ate it for you. The idea was shockingly simple: ingest tapeworm eggs, let the parasite feast on your food, and watch the pounds melt away. Advertisements even promised "safe" pills for expelling the worm when you reached your desired weight, although success stories were likely as scarce as they were dubious.

Other extreme measures included wearing vibrating belts or relying on bizarre machines that promised to "jiggle" fat away. Combined with tight corsetry and toxic tonics, Victorian weight-loss trends were a veritable buffet of discomfort and danger. Although their methods often failed spectacularly, they serve as a reminder that the quest for quick fixes is nothing new—and that humans will endure almost anything to win the battle of the bulge, parasites and all.

The 20th Century: Grapefruit, Cigarettes, and Jazzercise

With the steady discoveries of science, the 20th century saw a boom in fad diets, each promising rapid weight loss with its own quirky rules. Among the most famous was the Grapefruit Diet, also known as the "Hollywood Diet," which claimed that eating half a grapefruit before every meal would magically burn fat. Although popular, it offered little more than monotony and citrus overload.

Low-carb crazes also made their mark, starting with the 1950s Atkins-inspired plans that demonized bread and pasta while glorifying bacon and butter. These diets cycled in and out of popularity, with variations resurfacing every few decades, often wrapped in slicker branding.

From calorie-counting to bizarre food combinations, these fads reflected the growing cultural obsession with weight loss combined with interpretations of early research. Some had kernels of truth, but most fizzled out, leaving dieters hungry for the next big thing—preferably one with fewer grapefruits involved.

The 20th century's weight-loss trends were heavily shaped by marketing and media, which turned slimming down into a cultural phenomenon. Cigarette companies shamelessly capitalized on the thin-is-in mindset, with ads like Figure 2-6 claiming products like Lucky Strike could "keep you slender" by curbing appetite—a dangerous ploy that linked health to addiction.

Figure 2-6: Lucky Strike advertisement

Source: American Tobacco Company / Women in Tobacco Advertisements 1929-1939 / Public Domain

Meanwhile, celebrities jumped on the fitness bandwagon, popularizing programs like Jane Fonda's aerobics videos and Richard Simmons' enthusiastic weight-loss campaigns. Their charisma and media reach made fitness seem accessible and glamorous, drawing millions into Jazzercise classes and home workouts.

Diet product marketing flourished, too, with everything from meal-replacement shakes to questionable pills promising "effortless" results. These campaigns not only fueled the weight-loss industry but also reinforced societal pressures for thinness, setting the stage for the modern, media-driven obsession with dieting and body image.

The 20th century saw the rise of structured weight-loss plans like Weight Watchers, which brought a methodical, group-support approach to slimming down—along with a lot of calorie-counting guilt. Although some embraced the camaraderie and point-based system, others grumbled about trading brownies for bland rice cakes. Meanwhile, diet pills hit the market with promises of instant results, often conveniently glossing over side effects like jitters, insomnia, or outright heart palpitations.

Some pills relied on questionable ingredients, from amphetamines to thyroid hormones, essentially revving up metabolisms—and anxiety.

Continues

(Continued)

Ads boasted claims of shedding "10 pounds in a week," although what disappeared was often water weight, not fat. These plans and pills marked a shift toward more formulaic approaches to weight loss, some genuinely helpful, others laughably unrealistic. Still, they fueled an industry that thrived on the hope (and occasional desperation) of those chasing the ever-elusive "ideal" body.

The Modern Era: Science Meets Slimming

In the modern era, weight management has shifted from fad-driven approaches to science-backed strategies, thanks to advances in technology and biology. Wearables like fitness trackers and smartwatches now provide real-time data on activity, calories burned, and even sleep quality, empowering users to make informed choices. Meanwhile, breakthroughs in genetics and metabolism have revealed the complexity of weight regulation, moving beyond simple "calories in, calories out" models.

Innovations like personalized nutrition plans, metabolic testing, and gut microbiome analysis are transforming how individuals approach weight loss, tailoring solutions to their unique biology. Even medications, like GLP-1 receptor agonists such as Ozempic, Wegovy, Zepbound, and Mounjaro, are offering new hope for those struggling with obesity, targeting appetite and metabolism at a biochemical level.

Science has brought a nuanced understanding of weight, emphasizing sustainability and health over quick fixes. The struggle isn't over, but modern tools are bringing us closer to effective, personalized solutions.

The modern era has seen a shift from fad diets to scientifically grounded weight-management methods, prioritizing evidence-based solutions over gimmicks. Calorie-tracking apps, like MyFitnessPal, have revolutionized dieting by offering real-time insights into food intake and activity (see Figure 2-7), empowering users with data rather than restrictive food lists. These tools encourage sustainable habits, moving away from the all-or-nothing mindset of older fads.

For individuals facing severe obesity, bariatric surgery offers a transformative, medically proven solution. Procedures like gastric bypass and sleeve gastrectomy not only aid in weight loss but also address related conditions like diabetes and hypertension, demonstrating the effectiveness of a science-driven approach.

This shift reflects a growing recognition that weight management is multifaceted, requiring personalized, long-term strategies rather than quick fixes. Although the allure of miracle cures persists, the modern focus is on health, sustainability, and leveraging technology for lasting success.

Today's weight-loss culture is as much about hashtags as it is about health, with keto memes and fitness influencers dominating social media feeds. Who hasn't seen a post claiming that cauliflower crust pizza will "change your life" or a keto devotee proudly declaring, "Bread is the enemy!"? Meanwhile, fitness influencers flood timelines with impossibly toned selfies, accompanied by dubious advice like "Just drink this detox tea and watch the pounds melt away!"

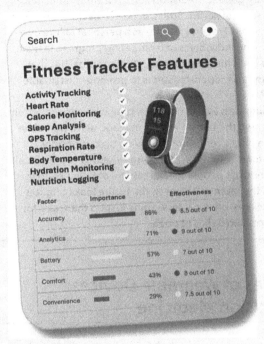

Figure 2-7: Fitness tracker features
Source: Generated with AI using DALL·E - OpenAI

Humor aside, the proliferation of misinformation is no joke. Quick-fix trends and unverified claims often overshadow genuine science, making it harder for people to separate fact from fiction. Yet the rise of online communities also fosters motivation and connection. As memes remind us, "You don't need a six-pack to be happy—but it wouldn't hurt if you still had tacos!" It's a quirky mix of inspiration and confusion in the digital age of weight management.

The journey of weight management has been a wild ride, from ancient remedies like camel milk and sweat baths to snake oil tonics and tapeworm diets. Over the centuries, humanity's efforts to slim down have ranged from bizarre to ingenious, reflecting both our creativity and desperation. Today, we stand at the intersection of science and technology, where wearables, personalized nutrition, and precision medicine are transforming how we approach weight and health. Lessons from the past remind us to be skeptical of quick fixes while embracing hope for a future rooted in evidence and innovation.

Continues

(Continued)

Understanding obesity as a complex, multifactorial challenge is essential for fostering effective solutions and compassion. Obesity isn't merely a result of willpower; it's influenced by genetic predispositions, hormonal imbalances, environmental factors, and societal norms. Stress, access to healthy food, and even sleep patterns all play a role. Framing obesity as a personal failure oversimplifies the issue, stigmatizing individuals and hindering progress. Instead, recognizing its complexity allows for tailored, empathetic approaches that integrate science, technology, and support systems to promote sustainable health improvements for all.

After centuries of grappling with everything from tapeworms to grapefruit, humanity may finally be turning the tide on weight management. Wearables and personalized healthcare offer a refreshing shift—no bizarre potions required! By providing real-time insights and tailoring solutions to individual needs, these innovations empower people to take control of their health in meaningful ways. Although the journey isn't over, the future looks brighter (and a lot less weird). With science on our side, we may just tip the scales for good—and still enjoy the occasional slice of pizza.

Costs and Outcomes of Wearable Devices: Traditional vs. Precision Medicine

Wearable devices have emerged as pivotal tools in managing chronic conditions, bridging the gap between traditional and precision medicine. By continuously tracking vital metrics such as heart rate, glucose levels, and physical activity, wearables provide real-time insights that empower both patients and healthcare providers. Unlike traditional medicine's episodic approach, wearables enable ongoing monitoring, fostering early detection and proactive management of conditions like diabetes and hypertension. Integrated with precision medicine, these devices contribute to personalized care plans, improving outcomes, enhancing engagement, and reducing the long-term costs associated with chronic disease management.

Traditional medicine often operates reactively, addressing health issues only after symptoms appear and relying on generalized treatment protocols. This one-size-fits-all approach, although effective for acute conditions, can fall short in managing chronic diseases, where personalized interventions are crucial. In contrast, precision medicine takes a proactive stance, tailoring care to an individual's unique genetic, environmental, and lifestyle factors. By focusing on prevention and early intervention, precision medicine not only improves outcomes but also minimizes unnecessary treatments. This shift marks a significant evolution toward more effective, patient-centered healthcare.

Wearable devices are reshaping costs and outcomes in healthcare, particularly for chronic conditions like obesity, by bridging traditional and precision medicine. In traditional, reactive paradigms, the lack of continuous monitoring often leads to costly interventions once conditions worsen. Wearables, however, provide real-time data that enables early detection and personalized management, aligning with precision medicine's proactive approach. By empowering patients to monitor activity, calorie expenditure, and sleep patterns, these devices reduce reliance on expensive treatments and improve outcomes. Their integration marks a pivotal step toward cost-effective, individualized care for chronic conditions.

Wearables and Traditional Medicine

In traditional medicine, wearable devices and their basic metrics—like steps taken, caloric expenditure, and sleep patterns—are often underutilized despite their potential to aid chronic condition management. Physicians in this model typically rely on episodic patient visits and lab tests for insights, leaving wearable-generated data as an optional, patient-driven tool rather than a core component of care. This disconnect limits opportunities for real-time, data-informed interventions.

Additionally, traditional medicine frequently places the financial burden of wearables on the patient. Devices are often considered lifestyle tools rather than essential medical equipment, meaning they're rarely covered by insurance. This cost barrier can exclude lower-income patients, reducing the overall accessibility and effectiveness of wearable-assisted care. Wearables can enhance traditional care by offering continuous monitoring and early insights, but their integration remains sporadic, reflecting a gap in leveraging technology for holistic chronic disease management.

Wearable devices, particularly entry-level models like basic fitness trackers, are relatively affordable and accessible, with prices often starting under $100. These devices (see Table 2-1) provide valuable metrics such as steps, heart rate, and calorie estimates, offering a cost-effective introduction to health monitoring. However, their limited functionality and generalized data can lead to inefficiencies in managing chronic conditions.

Table 2-1: Sample fitness tracker options

DEVICE	KEY FEATURES	PRICE
Amazfit Band 7	Heart rate monitoring, sleep tracking, 18-day battery life	$50
FITVII	Blood pressure monitoring, heart rate tracking, sleep tracking	$50
Fitbit Inspire 3	Heart rate monitoring, sleep tracking, 10-day battery life	$100
POLAR Unite	Heart rate monitoring, sleep and recovery tracking	$160
Garmin Vivoactive 5	Advanced health metrics, built-in GPS, music storage, multiday battery life	$200
WHOOP 4.0	Continuous health monitoring, strain and recovery tracking, subscription-based pricing	$200 (plus subscription)
Samsung Galaxy Watch 7	Advanced fitness features, strong battery life, integration with Samsung devices	$240
Oura Ring 4	Sleep and readiness tracking, heart rate monitoring, discreet design	$350
Apple Watch Series 10	Enhanced health metrics, sleep apnea detection, integration with iOS devices	$360

Source: Adapted from Fitness Trackers, December 28, 2024.

Without personalization—such as integrating wearable data with a patient's medical history, genetics, or specific health goals—users may struggle to interpret results or take meaningful action. For example, tracking steps is helpful, but without understanding how activity levels correlate with a condition like diabetes or obesity, the data may not translate into improved outcomes. These inefficiencies highlight the need for a more tailored approach, where wearable devices are not only affordable but also integrated into personalized care plans, ensuring that their full potential is realized in both cost-effectiveness and health impact.

Wearable devices often yield modest short-term improvements in weight management and chronic disease indicators, such as increased physical activity, initial weight loss, and better blood sugar control. Users frequently feel motivated by real-time feedback, setting goals and making healthier choices in the short run. These gains demonstrate the potential of wearables to enhance self-awareness and promote behavior change.

However, their impact on underlying causes or long-term success is more limited. Many users struggle to sustain initial progress, as wearable data alone may not address deeper challenges such as genetic predispositions, environmental factors, and psychological barriers. Additionally, without professional guidance or integration into comprehensive care plans, wearable-driven efforts often lack the tailored strategies needed for lasting results.

Although wearables are a promising tool for initiating change, their true potential lies in combining them with personalized interventions and support systems that tackle the root causes of chronic conditions.

Reliance on generalized protocols and wearable data often leads to inconsistent patient outcomes, particularly when the data lacks personalization. For example, a wearable device might recommend a generic daily step goal, like 10,000 steps, which can be too ambitious for someone with limited mobility or insufficient for a highly active individual managing obesity. This one-size-fits-all approach may demotivate patients who cannot meet the target or fail to optimize outcomes for those who can exceed it.

Similarly, wearables tracking caloric expenditure often use generalized formulas based on weight and activity, which may not account for metabolic differences. A patient with a slower metabolism may overestimate the calories they can consume, undermining weight-loss efforts. These inconsistencies arise when wearable data isn't contextualized with medical, genetic, or lifestyle factors. Without tailoring protocols to individual needs, the benefits of wearable-assisted care remain uneven, highlighting the need for more integrated, precision-focused solutions.

Wearables and Precision Medicine

Precision medicine leverages advanced wearable metrics to create highly individualized care plans for chronic condition management, transforming wearable devices from basic tracking tools into integral components of healthcare. Devices like CGMs offer real-time data on blood sugar levels, enabling precise adjustments in medication, diet, and activity for individuals managing diabetes. Unlike traditional episodic care, these wearables allow for dynamic interventions based on minute-to-minute variations, improving outcomes and reducing complications.

Similarly, sleep quality analysis from advanced wearables can inform precision care plans for conditions like obesity or cardiovascular disease. Poor sleep is a known contributor to metabolic dysfunction and weight gain, and wearables help identify disruptions, guiding interventions such as improved sleep hygiene or targeted therapies.

Metabolic data, such as energy expenditure and recovery rates, further enriches care by tailoring exercise and nutrition plans to an individual's unique physiology. This integration ensures that wearable data directly informs treatment, creating a feedback loop that enhances effectiveness, optimizes costs, and promotes sustainable health outcomes in ways traditional methods cannot achieve.

NOTE   Wearable devices and precision medicine get in front of outcomes and costs, not behind them.

Advanced wearable devices like CGMs, metabolic trackers, and advanced sleep monitors come with higher upfront costs compared to basic models. These devices often require significant investment, ranging from hundreds to thousands of dollars, and may include subscription fees for data analysis platforms. Additionally, ongoing interpretation of the complex data they generate often necessitates involvement from healthcare professionals, further adding to costs.

However, these initial expenses are offset by substantial potential savings over time. Advanced devices enable precise, real-time monitoring and early intervention, reducing the likelihood of complications from chronic conditions like diabetes, hypertension, or obesity. For instance, a CGM can prevent severe hyperglycemia or hypoglycemia, avoiding costly emergency visits or hospitalizations. Similarly, insights into sleep and metabolic health help address root causes, decreasing reliance on medications and preventing the progression of related conditions.

By enabling proactive, data-driven care, advanced wearables reduce the need for expensive reactive treatments, making them a cost-effective solution for managing chronic diseases in the long term (see Table 2-2).

Table 2-2: Wearable device outcomes

WEARABLE DEVICE	FUNCTION	KEY BENEFIT	TYPICAL OUTCOME
Continuous glucose monitor (Sensor patch)	Real-time blood glucose tracking	Understanding of glucose patterns	HbA1c reduction: 0.3%–1.2% Hypoglycemic time-in-range reduction: 30%–40%[4]
Blood pressure monitor (Band)	Continuous blood pressure measurement	Hypertension management, medication adherence	Systolic/diastolic reduction: 4–21 mm Hg[5]
Wearable ECG device (Smartwatch, chest strap)	Continuous cardiac rhythm monitoring	Early detection of arrhythmia	AFib episode detection: 70%–99% accuracy[6]
Oxygen saturation monitor (Smartwatch, ring, clip-on sensor)	SpO2 level monitoring, especially for chronic conditions	Management of respiratory or cardiac conditions	SpO2 predictive reliability: 83%–98%[7]
Cardiovascular tracker (Smartwatch, chest strap)	Heart rate, heart rate variability, arrhythmia detection	Stress and cardiovascular awareness	Diastolic reduction: 1–2 mm Hg HRV improvement: 8%–60%[8]
Sleep monitor (Ring, band, smartwatch)	Sleep stage analysis, detecting disturbances	Enhanced sleep quality, reduced fatigue	Sleep latency reduction: 48%–74%[9]
Metabolic tracker (Band, smartwatch)	Energy expenditure, activity levels, recovery rates	Motivation for exercise, meal planning	BMI decrease: 16%–19%[10]

Continues

Table 2-2 (continued)

WEARABLE DEVICE	FUNCTION	KEY BENEFIT	TYPICAL OUTCOME
Fitness and activity tracker (Band, smartwatch)	Steps, calories burned, workout tracking	Engagement in physical activity	Increased steps: 1,850 steps per day MET increase: 2.0–5.5 METs[11]
Stress and mental health monitor (Ring, smartwatch)	Stress level detection, biofeedback	Stress management, emotional awareness	HRV improvement: 22%–64% Breathing rate decrease: 19.1%[12]

The integration of wearables into precision medicine yields significant improvements in patient outcomes by enhancing adherence, personalizing interventions, and preventing chronic condition progression. Real-time feedback from CGMs or advanced fitness trackers fosters greater engagement, as patients can immediately see the impact of their actions on their health metrics. This visibility motivates adherence to care plans, such as sticking to prescribed exercise routines or adjusting dietary habits.

Wearables also enable highly personalized interventions. For example, a patient with diabetes can use CGM data to fine-tune insulin doses or make real-time adjustments to meals, resulting in better glycemic control. Similarly, wearables tracking sleep or stress provide actionable insights that guide tailored therapies for improving metabolic and cardiovascular health.

By addressing risk factors early and monitoring health continuously, wearables in precision medicine prevent chronic conditions from worsening. These measurable health gains—such as reduced blood pressure, stabilized blood sugar, and sustainable weight loss—highlight the transformative potential of wearable-assisted precision care in improving quality of life and reducing long-term healthcare costs.

The Bottom Line

The economic implications of wearable use differ significantly between traditional and precision medicine. Traditional medicine often incurs lower initial costs, as it typically underutilizes advanced wearable metrics and infrastructure. Patients frequently pay for basic devices out of pocket, and their data is rarely integrated into clinical care. Although this approach minimizes upfront expenses, it often results in higher long-term costs due to inefficiencies. The lack of continuous monitoring and tailored interventions can lead to recurrent treatment needs, avoidable hospitalizations, and the progression of chronic conditions, driving up healthcare expenditures.

In contrast, precision medicine involves higher initial investments in advanced wearables, data integration platforms, and professional expertise for interpretation. However, this proactive approach leverages wearables to deliver personalized, preventive care, targeting root causes and mitigating risks before complications arise. By reducing hospital admissions, improving adherence, and optimizing treatment plans, precision medicine has the potential to achieve significant long-term cost savings. For example, better glycemic control from continuous glucose monitoring reduces diabetes-related complications, and tailored sleep interventions decrease cardiovascular risks.

Overall, although precision medicine requires substantial upfront funding, its focus on prevention and individualized care positions it as a cost-effective, sustainable solution for managing chronic conditions with wearables (see Table 2-3).

Table 2-3: Costs and outcomes of wearable devices in treatment of chronic conditions

CHRONIC CONDITION	AFFECTED PATIENTS (UNITED STATES)	WEARABLE DEVICE SOLUTION	AVERAGE DEVICE COST	AVOIDABLE MEDICAL COST (PPPY)	PRECISION MEDICINE VALUE
Cancer[13]	18.1 million	Skin monitor, activity tracker	$650	$11,713	Very high
Cardiovascular Disease[14]	121.5 million	Heart rate monitor, blood pressure monitor	$800	$1,521	Very high
Diabetes[15]	38.1 million	Continuous glucose monitor, activity tracker	$3,700	$4,347	High
Digestive Disease[16]	62.0 million	Hydration tracker, stress monitor	$375	$2,414	High

Continues

Table 2-3 (*continued*)

CHRONIC CONDITION	AFFECTED PATIENTS (UNITED STATES)	WEARABLE DEVICE SOLUTION	AVERAGE DEVICE COST	AVOIDABLE MEDICAL COST (PPPY)	PRECISION MEDICINE VALUE
Behavioral Health[17]	59.3 million	Stress monitor	$300	$1,738	Moderate
Chronic Respiratory Disease[18]	33.8 million	Pulse oximeter, air quality monitor	$325	$1,175	Moderate

Note: Data is from the most recent available year. Avoidable medical cost is the average medical cost per patient per year (PPPY) less average device cost; precision medicine value is an approximate categorization based on the number of affected patients times the avoidable medical cost.

Wearable devices, when integrated into the framework of precision medicine, serve as powerful catalysts for transforming healthcare. Their ability to continuously collect real-time data on metrics such as activity levels, heart rate, glucose levels, and sleep quality enables a shift from reactive, episodic care to proactive, personalized health management. This synergy enhances early detection, prevention, and tailored treatment of chronic conditions like diabetes, hypertension, and obesity. This enables highly individualized interventions, fostering patient engagement by providing immediate feedback and encouraging adherence to care plans and lifestyle changes.

These results at the individual level build to comprehensive transformation by generating valuable population health data. This informs research, improves resource allocation, and guides public health policies. When coupled with precision medicine's targeted approach, wearables bridge the gap between innovation and accessibility, reducing healthcare costs through prevention and efficient chronic disease management. Their role in this framework positions them as pivotal tools for advancing a more dynamic, accessible, and effective healthcare system.

As technology evolves, wearables will become even more accessible and sophisticated, integrating seamlessly with healthcare ecosystems to promote rational, data-driven solutions. Their ability to foster engagement and deliver actionable insights positions them as key tools in combating chronic diseases, improving quality of life, and reducing healthcare costs. Looking ahead, wearable technology will play a critical role in shaping a future where healthcare is more effective, personalized, and universally accessible.

What You Can Do

As we have seen, wearable devices are powerful, accessible tools for monitoring and managing personal health, offering real-time insights into metrics like activity levels, sleep quality, and heart rate. These devices empower us to take an active role in our well-being, providing the information we need to make informed decisions about lifestyle and health choices. When used in collaboration with our healthcare providers, wearables can significantly enhance our health outcomes, improving personalized care plans and the results from proactive intervention.

Depending on your individual health circumstances—the foundation of precision medicine—there are several actions you can consider implementing in your life today to improve your health:

- *High blood pressure (hypertension).* Hypertension, or high blood pressure, affects nearly half of all adults, making it a leading risk factor for heart disease and stroke.[19] Wearable blood pressure monitors, like the Omron HeartGuide and Aktiia Bracelet, track blood pressure trends throughout the day. These devices sync with applications like Omron Connect and Aktiia's app to provide real-time insights. By identifying patterns, such as consistently elevated readings, wearables can inform healthcare professionals, enabling medication adjustments and lifestyle changes, such as improved diet, increased physical activity, and stress management. This proactive approach can help you achieve better long-term blood pressure control.

- *Cardiovascular health (arrhythmia, stroke).* Advanced cardiovascular risk affects approximately 51% of adults due to factors like high blood pressure, obesity, and inactivity.[20] Smartwatches like the Apple Watch and Fitbit Sense, along with heart rate monitors such as the Polar H10, track heart rate and variability. Paired with apps like Apple Health and Fitbit's platform, these devices can detect arrhythmias like atrial fibrillation, a major stroke risk, and provide insights into heart rate variability, an indicator of stress and cardiovascular health. These tools enable early detection and, with your healthcare professional, can guide interventions and support personalized strategies for you to manage and reduce cardiovascular risk.

- *Weight management and fitness (weight gain, fatigue, stress management).* Obesity affects approximately 42% of adults, posing risks for diabetes, cardiovascular disease, and other health conditions.[21]

Wearable devices like Fitbit, Garmin smartwatches, and the WHOOP strap, along with activity trackers and calorie counters, monitor movement, exercise, and caloric intake. Paired with apps like MyFitnessPal, Fitbit, and Garmin Connect, these devices encourage activity, track workout progress, and provide nutritional guidance. By offering personalized insights, these devices can help you set realistic goals, make informed dietary choices, and sustain weight loss or maintenance, empowering you to manage your weight and fitness effectively.

- *Sleep disorders (restlessness, insomnia, sleep apnea).* Sleep disorders affect approximately one-third of adults, with conditions like insomnia, sleep apnea, and restless leg syndrome disrupting health and daily life.[22] Wearables like the Oura Ring, Fitbit Sense, and Garmin smartwatches, along with sleep trackers such as Withings Sleep Analyzer, monitor sleep stages and disturbances. Apps like Oura, Fitbit, and Garmin Connect provide detailed sleep insights, flagging irregularities such as reduced deep sleep or interrupted breathing. These results can help you identify potential issues like sleep apnea or insomnia, guiding you to seek further evaluation with your healthcare provider and enabling better sleep hygiene practices.

- *Nutrition and blood sugar control (prediabetes, eating disorders).* Nutrition and blood sugar disorders, including diabetes, affect approximately 11% of adults, with many more experiencing pre-diabetes.[23] Continuous glucose monitors like the Dexcom G6, FreeStyle Libre, and Eversense provide real-time blood sugar tracking. These devices sync with apps such as Dexcom Clarity, LibreLink, and Eversense Mobile, offering detailed insights into glucose levels and trends. By enabling precise insulin management and dietary adjustments, CGMs can empower you to stabilize blood sugar, reduce complications, and make informed dietary and lifestyle choices to improve your overall health and nutrition.

When integrating wearables like these into your health routine, start by identifying the health metrics most relevant to your goals or conditions. For weight management, prioritize devices that track activity, calories burned, and sleep patterns, such as Fitbit and Garmin. If managing a condition like hypertension or diabetes, consider specialized devices like Omron HeartGuide for blood pressure and Dexcom G6 for continuous glucose monitoring.

Choose wearables that fit your lifestyle and offer proven accuracy. For example, sleek smartwatches may suit a tech-savvy user, and durable fitness trackers cater to active lifestyles. Look for ease of use, with intuitive apps that provide actionable insights and integrate with other health tools. Prioritizing these factors ensures that your wearable supports sustainable progress toward your health goals while seamlessly fitting into your daily routine.

It's important, too, that you collaborate with your healthcare provider, to maximize the benefits of your wearable devices. Start by discussing your health goals—whether managing weight, controlling blood pressure, or improving sleep—and set practical, achievable objectives. Your provider can help interpret your wearable data and integrate it into your overall care plan, ensuring that you focus on metrics that matter most for your health.

Consistent monitoring is also key. Review your device's feedback regularly, such as activity levels, heart rate trends, and sleep quality, to track progress and identify areas for improvement. Being aware of this data keeps you engaged and motivated while providing valuable information to share with your provider during follow-ups.

Finally, use your wearable's insights to guide lifestyle changes. Adjust your diet by tracking calorie intake and nutrient balance, or modify your exercise routine based on activity and recovery data. If your wearable flags issues like poor sleep, explore relaxation techniques or adjust bedtime habits. By combining wearable data, professional guidance, and actionable changes, you can create a holistic approach to achieving lasting health improvements.

Embracing wearable technology as an ally in your daily health journey can empower you with real-time insights for proactive management. By tracking key metrics and making informed adjustments, you can take charge of your well-being, and by combining your personal efforts with professional guidance to align your wearable data with tailored care, these tools can create a path to lasting health improvements and your most vibrant and fulfilling life.

Notes

1. Wang, Lili, Si, Lei, Cocker, Fiona, Palmer, Andrew, and Sanderson, Kristy, "A Systematic Review of Cost-of-Illness Studies of Multimorbidity," *Applied Health Economics and Health Policy* 16, August (2018): 15–19.

Makovski, Tatjana, Schmitz, Susanne, Zeegers, Maurice, Stranges, Saverio, and van den Akker, Marjan, "Multimorbidity and Quality of Life: Systematic Literature Review and Meta-analysis," *Ageing Research Reviews* 53, August (2019): 100903.

World Health Organization, "Noncommunicable Diseases," September 16 (2023), https://www.who.int/news-room/fact-sheets/detail/noncommunicable-diseases.

2. Hajat, Cother, and Stein, Emma, "The Global Burden of Multiple Chronic Conditions: A Narrative Review," *Preventive Medicine Reports* 12, October 19 (2018): 284–293.

3. Our World in Data, "Annual Number of Deaths by Cause," (2024), https://ourworldindata.org/grapher/annual-number-of-deaths-by-cause.

4. Association of Diabetes Care & Education Specialists, "Effectiveness of CGMs," December 30 (2024), https://www.adces.org/education/danatech/glucose-monitoring/continuous-glucose-monitors-%28cgm%29/cgm-101/effectiveness-of-cgms.

5. Gazit, Tomer, Gutman, Michal, and Beatty, Alexis, "Assessment of Hypertension Control Among Adults Participating in a Mobile Technology Blood Pressure Self-management Program," *JAMA Network Open* 4, no. 10 (2021): e2127008.

6. Zarak, Muhammed, Khan, Sher, Majeed, Harris, Yasinzai, Abdul, Hamzazai, Wadana, Chung, Duy, Koshkarian, Gregory, and Fleming, Kevin, "Systematic Review of Validation Studies for the Use of Wearable Smartwatches in the Screening of Atrial Fibrillation," *International Journal of Arrhythmia* 25, no. 11 (2024).

7. Walzel, Simon, Mikus, Radek, Rafl-Huttova, Veronika, Rozanek, Martin, Bachman, Thomas, and Rafl, Jakub, "Evaluation of Leading Smartwatches for the Detection of Hypoxemia: Comparison to Reference Oximeter," *Sensors* 23, no. 22 (2023): 9164.

8. Bolin, Linda, Saul, Amelia, Bethune Scroggs, Lauren, and Horne, Carolyn, "A Pilot Study Investigating the Relationship between Heart Rate Variability and Blood Pressure in Young Adults at Risk for Cardiovascular Disease," *Clinical Hypertension* 28, no. 2 (2022).

9. Elemind, "Elemind Launches First-of-its-Kind 'Sleep, On-Demand' Headband, Giving Control to Millions Struggling with Sleep," June 4 (2024), https://www.businesswire.com/news/home/20240604568537/en.

10. Keskin, Bülent, and Köse, Esra, "The Effect of Calorie Tracking on Eating Behaviors of University Students," *Journal of Science Learning* 6, no. 2 (2023): 216–221.

11. Laranjo, Liliana, Ding, Ding, Heleno, Bruno, Kocaballi, Baki, Quiroz, Juan, Tong, Huong, Chahwan, Bahia, Neves, Ana, Gabarron, Elia, Dao, Kim, Rodrigues, David, Neves, Gisela, Antunes, Maria, Coiera, Enrico, and Bates, David, "Do Smartphone Applications and Activity Trackers Increase Physical Activity in Adults? Systematic Review, Meta-Analysis and Metaregression," *British Journal of Sports Medicine* 55 (2021): 422–432.

12. Lin, I-Mei, Chen, Ting-Chun, Tsai, Hsin-Yi, and Fan, Sheng-Yu, "Four Sessions of Combining Wearable Devices and Heart Rate Variability (HRV) Biofeedback Are Needed to Increase HRV Indices and Decrease Breathing Rates," *Applied Psychophysiology and Biofeedback* 48, no. 2 (2022).

13. National Cancer Institute, "Statistics on Cancer Survivorship," August 1 (2024), https://cancercontrol.cancer.gov/ocs/statistics.

    National Cancer Institute, "Cancer Statistics," May 9 (2024), https://www.cancer.gov/about-cancer/understanding/statistics.

14. American Heart Association, "2024 Heart Disease and Stroke Statistics: At-a-Glance," (2024), https://www.heart.org/-/media/PHD-Files-2/Science-News/2/2024-Heart-and-Stroke-Stat-Update/2024-Statistics-At-A-Glance-final_2024.pdf.

    American Heart Association, "Cardiovascular Diseases Affect Nearly Half of American Adults, Statistics Show," January 31 (2019), https://www.heart.org/en/news/2019/01/31/cardiovascular-diseases-affect-nearly-half-of-american-adults-statistics-show.

15. Centers for Disease Control and Prevention, "Data & Research | Diabetes," May 15 (2024), https://www.cdc.gov/diabetes/php/data-research/index.html.

    American Diabetes Association, "New American Diabetes Association Report Finds Annual Costs of Diabetes," November 1 (2023), https://diabetes.org/newsroom/press-releases/new-american-diabetes-association-report-finds-annual-costs-diabetes-be.

16. GI Alliance, "Digestive Disease Continues to Rise Among Americans," February 23 (2021), https://gialliance.com/

`gastroenterology-blog/digestive-disease-continues-to-rise-among-americans.`

Peery, Anne, Crockett, Seth, Murphy, Caitlin, Lund, Jennifer, Dellon, Evan, Williams, Lucas, Jensen, Elizabeth, Shaheen, Nicholas, Barritt, Alfred, Lieber, Sarah, Kochar, Bharati, Barnes, Edward, Fan, Claire, Pate, Virginia, Galanko, Joseph, Baron, Todd, and Sandler, Robert, "Burden and Cost of Gastrointestinal, Liver, and Pancreatic Diseases in the United States: Update 2018," *Gastroenterology* 156, no. 1 (2018): 254–272.

17. National Institute of Mental Health, "Mental Illness," September (2024), `https://www.nimh.nih.gov/health/statistics/mental-illness.`

    Agency for Healthcare Research and Quality, "Medical Expenditure Panel Survey: Statistical Brief #539," February (2022), `https://meps.ahrq.gov/data_files/publications/st539/stat539.pdf.`

18. American Lung Association, "Methodology: Prevalence and Incidence of Lung Disease," June 7 (2024), `https://www.lung.org/research/trends-in-lung-disease/prevalence-incidence-lung-disease/methodology.`

    McKenna, Jon, "Costs of COPD," April 21 (2023), `https://www.webmd.com/lung/copd/costs-of-copd.`

19. Centers for Disease Control and Prevention, "Facts About Hypertension," May 15 (2024), `https://www.cdc.gov/high-blood-pressure/data-research/facts-stats/index.html.`

20. Joynt Maddox, Karen, Elkind, Mitchell, Aparicio, Hugo, Commodore-Mensah, Yvonne, de Ferranti, Sarah, Dowd, William, Hernandez, Adrian, Khavjou, Olga, Michos, Erin, Palaniappan, Latha, Penko, Joanne, Poudel, Remy, Roger, Véronique, and Kazi, Dhruv, "Forecasting the Burden of Cardiovascular Disease and Stroke in the United States Through 2050—Prevalence of Risk Factors and Disease: A Presidential Advisory From the American Heart Association," *Circulation* 150, no. 4 (2024): e65–e88.

21. Elmaleh-Sachs, Arielle, Schwartz, Jessica, Bramante, Carolyn, Nicklas, Jacinda, Gudzune, Kimberly, and Jay, Melanie, "Obesity Management in Adults: A Review," *JAMA* 330, no. 20 (2023): 2000–2015.

22. Holder, Sarah, and Narula, Navjot, "Common Sleep Disorders in Adults: Diagnosis and Management," *American Family Physician* 105, no. 4 (2022): 397–405.

23. American Diabetes Association, "Statistics About Diabetes," November 2 (2023), `https://diabetes.org/about-diabetes/statistics/about-diabetes`.

The Science of Hope

Precision medicine often seems complex to laypeople because it combines cutting-edge science with advanced technology, which can feel overwhelming and inaccessible. At its core, it tailors medical treatment to the individual based on unique factors like genetics, lifestyle, and environment. This level of customization contrasts sharply with the more familiar, one-size-fits-all approach of traditional medicine.

To simplify, we can think of precision medicine like finding the perfect key for a lock. Traditional methods might involve trying a set of keys until one fits, wasting time and effort. Precision medicine, on the other hand, identifies the exact key—the treatment—needed for your specific lock—your condition. Cancer therapies can now target specific genetic mutations, for instance, acting like a precise key to shut the disease down with fewer side effects.

By using tools like wearable devices and genetic testing, precision medicine transforms trial and error into a science-backed roadmap, making healthcare both more effective and personal. With relatable technology like fitness trackers, these advancements feel less like science fiction and more like a logical step toward better health.

Simplifying Precision Medicine

Precision medicine is an approach to healthcare that tailors treatment and prevention strategies to an individual's unique genetic, environmental, and lifestyle factors. Unlike the generic approach of traditional medicine, which applies the same treatment to all of us regardless of our differences, precision medicine offers a personalized path to better health.

Cancer care, for example, has historically utilized treatments like chemotherapy that act broadly, targeting all rapidly dividing cells and causing significant side effects. Precision medicine analyzes the genetic profile of a patient's tumor and selects targeted therapies that attack the cancer cells without harming healthy tissue—like choosing the perfect tool for a delicate task rather than using a mass-marketed hammer.

The same principle applies to conditions like heart disease or diabetes. By understanding a person's genetic predispositions and how their lifestyle affects their health, doctors can recommend treatments and interventions that align closely with the individual's biology, environment, and habits. Precision medicine isn't just treatment; it's healthcare that sees you as a unique individual.

Precision medicine stands on four transformative pillars:

- Wearable devices
- Pharmacogenomics
- Big Data analytics
- AI

These each play a vital role in personalizing healthcare and improving outcomes.

Together, they integrate data and technology to deliver healthcare that is predictive, preventive, and profoundly personalized, transforming lives and the future of medicine.

How It Works: From Data to Decisions

Precision medicine transforms a patient's unique characteristics into actionable insights for healthcare. This process involves collecting, analyzing, and applying data using advanced technologies, making personalized care possible. Figure 3-1 shows a graphical depiction of this process.

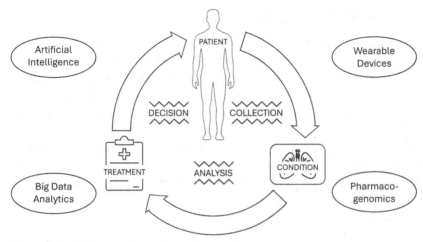

Figure 3-1: Clinical workflow of precision medicine

Step 1: Data Collection

The journey begins with gathering detailed information about a patient. Tools like genomic sequencing decode a person's DNA to identify genetic variations that can affect their health. For example, a genetic test might reveal whether someone is predisposed to certain cancers or how they might respond to a specific medication.

At the same time, wearable devices continuously monitor health metrics as they occur. Think of these as health trackers on steroids—they go beyond counting steps, measuring heart rate, blood pressure, glucose levels, and even sleep quality. This constant stream of data paints a comprehensive picture of a patient's daily health, capturing fluctuations that might otherwise go unnoticed in occasional doctor visits.

Step 2: Data Analysis

Once the data is collected, powerful tools are employed to make sense of it all. Big Data analytics compares an individual's data against patterns observed in millions of other patients, identifying risks and predicting outcomes. For instance, data from wearables and health records might highlight early warning signs of heart disease before symptoms appear.

Artificial intelligence acts like a digital assistant for doctors, processing these vast amounts of data quickly and accurately. AI algorithms can identify subtle trends, such as predicting how a specific treatment will affect a patient based on their unique genetic makeup and lifestyle.

Step 3: Turning Insights into Decisions

The final step is applying these insights to craft a tailored healthcare plan. For example, genomic sequencing may guide a doctor to prescribe a medication optimized for a patient's metabolism, avoiding side effects. Wearable data might suggest lifestyle adjustments like incorporating more activity to lower blood pressure or modifying mealtimes to stabilize glucose levels.

By combining genetic information, real-time health data, and advanced analytics, precision medicine replaces guesswork with evidence-based, individualized care. It empowers patients and providers to make informed decisions, ensuring healthcare is as dynamic and unique as the individuals it serves.

The Patient Perspective: Benefits Made Simple

In traditional medicine, patients often endure months or years of trial-and-error treatments. Precision medicine shortcuts this process. For example, genomic testing can reveal whether a specific cancer treatment will be effective, allowing doctors to target the disease immediately rather than wasting time on ineffective therapies.

Personalized care also minimizes unnecessary interventions and their associated costs. For instance, pharmacogenomics ensures that patients receive medications suited to their genetic makeup, reducing adverse reactions. This not only spares patients discomfort but also prevents additional medical expenses from addressing complications.

Finally, precision medicine provides patients with actionable insights about their health. With tools like wearable devices, individuals gain a clearer understanding of how lifestyle choices affect their well-being, enabling them to take proactive steps toward better health.

NOTE   Precision medicine empowers patients by addressing root causes of illness and infirmity instead of symptoms.

In this way, precision medicine doesn't just treat patients; it empowers them. By addressing the root causes of illness with precision and compassion, it offers hope and practical solutions, redefining the healthcare experience. And precision medicine isn't just for advanced medical treatments—it's becoming part of our daily routines, thanks to accessible technology like fitness trackers, wearable devices, and health apps.

These tools are bringing the benefits of precision medicine to everyday life, empowering people to monitor and improve their health in real time.

Take fitness trackers, for example. These devices do more than count steps—they monitor vital signs and activity levels, providing users with actionable insights. Imagine a fitness tracker alerting someone to irregular heart rhythms. This real-time warning could prompt them to visit a doctor, catching a condition like atrial fibrillation before it becomes a serious problem.

Continuous glucose monitors (CGMs) are another example. For individuals with diabetes, these devices provide constant updates on blood sugar levels, helping them make informed decisions about diet, exercise, and medication. No more guessing or waiting for periodic checkups—patients can take control of their health daily.

Even smartphone apps are making precision medicine accessible. Nutrition apps, for instance, analyze eating habits and suggest personalized meal plans based on goals like weight management or blood sugar control. Similarly, stress management apps use wearable data to guide mindfulness practices tailored to an individual's needs.

This integration of technology into everyday life makes precision medicine relatable and practical. By providing insights into health metrics and encouraging small, meaningful changes, these tools allow anyone—not just patients in a clinical setting—to benefit from personalized healthcare. Whether it's a smartwatch buzzing to remind you to move or a health app recommending better sleep habits, precision medicine is evolving into an indispensable part of daily living, empowering individuals to lead healthier, more informed lives.

Addressing Misconceptions

And yet, in its simplicity, precision medicine is often misunderstood, with myths and misconceptions overshadowing its potential. One common belief is that it's only for the wealthy—a luxury reserved for those who can afford cutting-edge technology or specialized treatments. Although precision medicine initially emerged from high-cost innovations like genomic sequencing, it is rapidly becoming more accessible and affordable. The cost of genetic testing, for example, has dropped significantly in the last decade, making it feasible for broader populations. In 2003, sequencing a human genome cost billions; today, it can be done for less than $1,000, with some targeted tests available for even less.[1] Many insurance providers are beginning to cover these tests, recognizing their value in preventing costly medical complications.

Another misconception is that precision medicine increases healthcare costs. In reality, it has the potential to save money by reducing waste and improving efficiency. Traditional healthcare often involves guessing at treatments, leading to prolonged illness, unnecessary interventions, and escalating expenses. Precision medicine eliminates much of this guesswork. For example, pharmacogenomics can help doctors prescribe the right medication the first time, avoiding the costs of ineffective treatments and managing side effects.

Some also believe precision medicine is limited to treating rare conditions or cancer. Although it has made remarkable strides in these areas, its applications are far broader. Everyday tools like health apps bring personalized care into the hands of millions. These technologies enable individuals to monitor their health, make lifestyle adjustments, and prevent issues before they arise—proof that precision medicine isn't confined to the elite.

As awareness grows and costs decline, precision medicine is moving closer to its ultimate goal: making personalized, effective care a standard for everyone, not a privilege for a few. Its emphasis on prevention and tailored treatment benefits patients and healthcare systems alike, proving that it's not just innovative—it's practical and essential for the future of healthcare.

A Vision for the Future

The future of precision medicine holds immense promise, with the potential to reshape healthcare into a more personalized, efficient, and equitable system. As advancements in technology and science continue to unfold, precision medicine is poised to become the cornerstone of healthcare, benefiting not just a few but the entire population.

Ongoing advancements are also making precision medicine more affordable and universal. The cost of genetic testing has plummeted, and healthcare systems are recognizing the economic value of investing in prevention and personalized care. AI and Big Data are streamlining the analysis of vast patient datasets, reducing the time and expense required to generate actionable insights. These innovations are not just for urban centers or wealthy nations—they're being adapted for use in rural and underserved communities, bringing the benefits of precision medicine to those who need it most.

Far from being reserved for researchers or specialists, precision medicine is steadily becoming a practical, everyday tool that empowers individuals to take charge of their health. Whether it's through tools that provide real-time insights, tests that guide personalized treatments, or diagnostics that simplify complex data, precision medicine is reshaping how we approach well-being.

At its heart, precision medicine is about recognizing and embracing what makes each of us unique—our genetics, lifestyles, and environments—and using that understanding to craft care that truly fits. It's about more than curing disease; it's about creating a system that values prevention, minimizes suffering, and optimizes health outcomes.

And the promise of precision medicine extends beyond individual patients. By reducing costs through targeted treatments and preventing complications before they arise, it also addresses systemic healthcare challenges, making care more efficient and equitable. This journey is not just for the scientifically inclined—it's a step toward a healthier future. By understanding and supporting the growing accessibility of precision medicine, we can all play a part in building a world where healthcare is personal, proactive, and universally accessible. It's a vision of hope, progress, and possibility—one that turns science into better health.

Understanding the Impact of Wearables, Pharmacogenomics, Big Data Analytics, and Artificial Intelligence

Precision medicine, once an aspiration, is becoming a reality thanks to breakthroughs in data, devices, and analysis across the spectrum of innovation. Together, these tools form the foundation of a new, personalized approach to health—one that is predictive, preventive, and tailored to the individual.

Individually, the innovations are impressive, but their synergy is what truly revolutionizes healthcare. Together, they enable doctors and patients to move beyond one-size-fits-all solutions, creating a seamless, personalized system of care. This integration of technology and medicine promises not only to improve outcomes but also to make healthcare more accessible and effective, ushering in a new era of innovation and hope.

Health in Real Time with Wearable Devices

Wearable devices are revolutionizing healthcare by bringing real-time health monitoring directly to individuals. Smartwatches can track daily activity and heart rate variability, giving users feedback on how exercise, diet, or stress affects their overall health, and CGMs offer invaluable data for individuals managing diabetes, enabling them to see how specific foods or activities impact their blood sugar levels in real time. Sleep trackers analyze sleep cycles, helping users optimize their rest for better recovery and mental clarity.

And emerging technologies promise even greater advancements. Devices capable of detecting biomarkers for diseases, such as cancer and Alzheimer's, are already in development. Smart clothing with embedded sensors can continuously monitor vital signs, and advanced patches can measure hydration, electrolyte levels, or even stress hormones. The potential for wearables extends beyond individual health management, too. Aggregated wearable data can inform broader public health initiatives, identifying trends and helping healthcare providers predict and prevent widespread issues.

By enabling individuals to monitor their health in real time, wearables transform vague aspirations for better health into measurable, actionable steps. They not only provide insights for improving day-to-day living but also serve as an early warning system for critical health issues, proving that precision medicine isn't confined to clinics—it's becoming an indispensable part of everyday life.

Matching Medicine to DNA Through Pharmacogenomics

Pharmacogenomics is the study of how a person's genetic makeup influences their response to medications. By understanding the genetic variations that affect drug metabolism, efficacy, and side effects, pharmacogenomics enables healthcare providers to tailor treatments specifically to an individual's needs. This personalized approach eliminates much of the trial and error typically associated with prescribing medications, leading to better outcomes and fewer adverse effects. As a case in point, patients with behavioral disorders like depression often endure months of trying different medications to find one that works, frequently experiencing debilitating side effects along the way. Pharmacogenomic testing can identify genetic markers that indicate how a patient will metabolize certain antidepressants, allowing doctors to prescribe the most effective medication from the start. This not only shortens the path to relief but also spares patients unnecessary discomfort.

In pain management, too, pharmacogenomics has shown remarkable promise. For instance, some patients metabolize opioids very rapidly, making these medications less effective and increasing the risk of side effects or addiction. Genetic testing can reveal such variations, guiding doctors to safer and more effective alternatives, such as non-opioid pain relievers or targeted therapies.

Beyond improving individual care, pharmacogenomics has significant economic implications. Traditional approaches to prescribing often lead to multiple failed treatments, prolonged illness, and additional medical costs. By identifying the right medication up front, pharmacogenomics reduces wasted resources and minimizes hospitalizations caused by adverse drug reactions. Patients with certain cancers can undergo genetic testing to determine whether a specific targeted therapy is likely to work, avoiding the expense and risks of treatments that might fail.

The benefits extend to healthcare systems as well. Pharmacogenomic insights can inform formularies and treatment protocols, ensuring that resources are directed toward the most effective solutions. By reducing unnecessary prescriptions, side effects, and follow-up care, pharmacogenomics contributes to a more efficient and cost-effective healthcare model. Pharmacogenomics exemplifies the power of precision medicine by aligning treatment with the unique biology of each patient. It not only improves patient outcomes but also streamlines healthcare delivery, proving that tailoring medicine to DNA is a game-changer for individuals and systems alike.

Harnessing the Power of Big Data

Big Data analytics is revolutionizing healthcare by aggregating vast amounts of information from patient records, wearable devices, genomic data, and other sources to uncover patterns and insights that were previously invisible. This wealth of information enables healthcare providers to move from reactive care to predictive, preventive, and highly personalized treatment.

Big Data has been instrumental in managing chronic diseases like diabetes, for example. By analyzing data from wearables and electronic health records (EHRs), algorithms can identify patterns such as blood sugar spikes linked to specific meals or activity levels. This allows doctors to provide tailored advice and helps patients take proactive steps to manage their condition. In one notable case, a medical device manufacturer used Big Data to predict hospital readmissions for heart failure patients, enabling early interventions that significantly reduced readmission rates.[2]

Another example comes from pandemic prediction models. Big Data has been used to track emergent infectious diseases by analyzing public health records, social media posts, and environmental factors. These models provide early warnings, helping governments and healthcare providers prepare resources and containment strategies.[3]

However, for Big Data analytics to achieve its full potential, effective integration and data privacy safeguards are essential. Healthcare systems must adopt interoperable platforms that seamlessly combine information from various sources. Equally important is ensuring the security of sensitive patient data. Robust encryption, anonymization, and clear consent policies are critical to building trust and protecting individual privacy. By identifying health risks, predicting trends, and tailoring treatments, Big Data analytics enhances precision medicine on both individual and population levels. It transforms information into actionable insights, allowing healthcare providers to intervene earlier, improve outcomes, and make care more efficient.

Artificial Intelligence: Healthcare's Digital Brain

AI is improving healthcare by serving as a "digital brain" capable of analyzing vast and complex datasets, diagnosing diseases with precision, and suggesting tailored treatment plans. Its ability to process and interpret information far exceeds human capacity, enabling healthcare providers to make more informed and effective decisions.

One of the most compelling applications of AI is in disease diagnosis. For instance, AI algorithms have been trained to analyze imaging data, such as mammograms, with remarkable accuracy.[4] In cancer care, these tools can detect subtle abnormalities that might be missed by human eyes, allowing for earlier and more precise diagnoses. AI-assisted pathology also helps identify the genetic markers of tumors, guiding oncologists toward the most effective targeted therapies.[5]

AI is also revolutionizing drug discovery. By analyzing millions of chemical compounds and their interactions with specific genes or proteins, AI accelerates the development of personalized medications. For example, AI tools have been used to identify potential treatments for rare diseases, significantly reducing the time and cost traditionally associated with drug development.[6]

Managing chronic conditions is yet another area where AI shines. Advanced algorithms can synthesize data from wearable devices, electronic health records, and patient-reported outcomes to provide real-time insights and recommendations. For a patient with diabetes,

AI might analyze glucose patterns to predict blood sugar fluctuations and suggest dietary adjustments or insulin doses, helping maintain stable levels and preventing complications.

NOTE   Although AI might identify a promising insight or approach, it's the doctor who implements the final treatment.

Importantly, AI is a tool to enhance, not replace, human decision-making. It provides clinicians with data-driven insights, empowering them to make better-informed choices while preserving the empathy and judgment that only humans can offer. Although AI might recommend a treatment plan based on statistical likelihoods, it's the doctor who considers the patient's preferences, lifestyle, and overall well-being before finalizing a decision.

Looking ahead, advanced AI applications like virtual health assistants and predictive healthcare hold immense promise. Virtual assistants can answer patient queries, schedule appointments, and provide reminders for medication adherence, improving accessibility and convenience. Predictive models powered by AI can forecast outbreaks of diseases or identify populations at risk for certain conditions, enabling preventive measures on a large scale. By streamlining processes, improving accuracy, and personalizing care, AI is becoming a reliable partner in modern healthcare.

Synergy Between Technologies

For all of the value of the individual innovations, the true power of precision medicine lies in the integration of wearable devices, pharmacogenomics, Big Data analytics, and AI. Together, these form a dynamic ecosystem that transforms raw health data into actionable, personalized care. This proactive approach doesn't just address immediate health issues—it prevents potentially life-threatening events and improves long-term health outcomes. By combining real-time monitoring, advanced analytics, and targeted treatment, the technologies work in harmony to provide care that is precise, effective, and timely.

The integration of wearables, pharmacogenomics, Big Data analytics, and AI represents a fundamental shift in healthcare. Together, these innovations are breaking down the barriers of traditional medicine, creating a world where care is proactive and not just generic but deeply personalized. By collecting real-time data, identifying patterns, and tailoring treatments to individual needs, this synergy enhances health outcomes while reducing inefficiencies. Imagine a future where illnesses are detected before symptoms appear, treatments are optimized for each person's unique genetic profile, and healthcare providers are equipped with tools to deliver precision care with unparalleled accuracy.

At the bottom line, these advancements are making healthcare more accessible. The falling costs of genetic testing, the ubiquity of wearable devices, and the scalability of AI-driven solutions ensure that precision medicine benefits not just a few but communities worldwide.

Macro-Level Cost and Outcomes

This shift to personalized care cannot come too soon. The escalating costs of traditional healthcare have reached a critical point, straining individuals, families, and economies worldwide. Chronic diseases, ineffective treatments, and avoidable complications contribute significantly to the financial burden. In the United States alone, healthcare expenditures exceed $4 trillion annually, with much of it coming from inefficiencies such as trial-and-error treatments, repeated hospitalizations, and delayed diagnoses.[7] This unsustainable trajectory demands innovative solutions.

Precision medicine offers the path forward. By leveraging tools like those we've identified, it moves healthcare from a reactive, generic model to a proactive, personalized approach. Precision medicine has the potential to minimize waste by matching treatments to patients' needs, reducing hospital stays, and preventing adverse drug reactions. It also focuses on prevention and early intervention, addressing health issues before they escalate into costly crises. However, adopting precision medicine on a large scale requires a closer examination of its broader financial and societal implications. Questions about accessibility, affordability, and infrastructure must be addressed to ensure that these advancements benefit everyone, not just a few.

As we explore the intersection of cost and outcomes, it becomes clear that precision medicine isn't just a scientific breakthrough—it's an opportunity to create a more efficient, equitable, and sustainable healthcare system. The potential savings and improved outcomes it offers represent a pathway to resolving the cost crisis and building a healthier future.

The High Cost of Trial-and-Error Medicine

As we have seen, the traditional healthcare model typically relies on generalized treatment methods, leading to significant costs for both patients and the systems that serve them. When treatments fail or produce adverse effects, the result is a cascade of repeated diagnostics, ineffective therapies, and prolonged hospital stays. This inefficiency not only strains healthcare resources but also imposes a heavy burden on patients, both emotionally and economically.

A stark example of this is that nearly 50% of patients with common conditions such as depression and hypertension do not respond to the first medication prescribed.[8] This failure often requires multiple cycles of alternative treatments, with each new attempt adding weeks or months to the process. During this time, patients often experience worsening symptoms, side effects, and mounting anxiety, all while incurring additional medical expenses such as follow-up consultations, lab tests, and hospitalizations.

One surprising glimpse of waste in the current system is the cost of adverse drug reactions, which are responsible for approximately $30 billion in annual direct medical expenses in the United States alone.[9] These events often result from prescribing medications without a clear understanding of how an individual's body will respond, leading to preventable complications and hospital readmissions.

The economic toll on patients is compounded by lost productivity, as extended treatment timelines often mean time away from work or reduced ability to perform daily activities. For families, the emotional impact of prolonged uncertainty and financial stress can be devastating. Precision medicine addresses these inefficiencies by tailoring treatments to the individual from the start, reducing the need for guesswork and minimizing waste. By identifying the most effective therapies up front, it holds the potential to alleviate the unnecessary financial and emotional burdens imposed by trial-and-error approaches, creating a more efficient and compassionate healthcare system. Patients not only recover faster but also avoid the expense of prolonged, generalized treatments.

Predictive analytics adds another opportunity for cost savings by enabling early detection and prevention. By analyzing wearable data, electronic health records, and other health indicators, predictive models can identify risks before symptoms manifest. Algorithms that detect early signs of heart failure or diabetes, for instance, allow for timely interventions, reducing the likelihood of expensive late-stage complications and hospitalizations.

And evidence supports these benefits. A study on pharmacogenomic testing showed a 15% reduction in hospital readmissions for patients whose treatments were guided by their genetic profiles.[10] Similarly, the use of AI-driven diagnostic tools in cancer care has been shown to save thousands of dollars per patient by reducing unnecessary tests and accelerating accurate diagnoses.[11]

By focusing on prevention, early detection, and tailored treatments, precision medicine streamlines healthcare, benefiting both patients and the broader system. These innovations not only improve outcomes but also promise more sustainable financial implications for healthcare.

Measuring Improved Outcomes

One key metric in improving outcomes is survival rates, particularly in cancer care. Targeted therapies based on genomic profiling, such as HER2 inhibitors for breast cancer and EGFR inhibitors for lung cancer, have significantly increased survival rates compared to traditional chemotherapy.

> **NOTE**   Human Epidermal Growth Factor Receptor 2 (HER2): Gene that produces a protein involved in cell growth and division. HER2 plays a critical role in regulating how cells grow, divide, and repair themselves.
>
> Epidermal Growth Factor Receptor (EGFR): Gene that encodes a protein crucial for cell growth, division, and survival. Mutations or overactivity of this gene can contribute to cancer.

For instance, studies show that targeted treatments for non-small-cell lung cancer improve survival rates by more than three times that of chemotherapy alone.[12]

Reduced side effects are another critical outcome. In analyses of real-world evidence, pharmacogenomic-guided medication selection has shown a 52% reduction in adverse drug reactions compared to conventional prescribing practices, enhancing patients' quality of life while reducing follow-up care needs.[13] Faster recovery times further demonstrate the impact of precision care. In chronic conditions like diabetes, wearable devices providing real-time glucose monitoring have helped patients maintain better blood sugar control, leading to fewer hospitalizations and quicker recovery from acute episodes. A study of patients using continuous glucose monitors found a 53% reduction in diabetes-related emergency room visits.[14]

These improved outcomes translate directly into economic benefits. Faster recovery and reduced complications mean fewer days off work, boosting productivity for both patients and their caregivers. In cancer care, where treatment often results in significant time away from professional and personal responsibilities, precision medicine has been shown to decrease treatment time and associated caregiver burdens by as much as 75%, reducing emotional and financial strain.[15]

Aggregated data highlights the broader impact: an analysis across multiple conditions found that precision medicine improved healthcare cost-effectiveness in 71% of studies, through fewer hospitalizations and more effective treatments.[16] Additionally, the adoption of precision-guided care in rare diseases has shortened the diagnostic odyssey—the time from symptom to diagnosis—from an average of more than five years to as little as five days, sparing patients years of uncertainty and unnecessary costs.[17]

Precision medicine's impact extends beyond individual patients, too, creating an economic ripple effect that benefits businesses, economies, and the broader healthcare landscape. By streamlining care delivery and eliminating ineffective practices, precision medicine reduces overall healthcare expenditure while fostering a healthier, more productive society.

At the system level, precision medicine saves costs by preventing complications, reducing hospitalizations, and optimizing treatments. Traditional generalized approaches often lead to wasted resources, including unnecessary tests, ineffective therapies, and prolonged hospital stays. Precision medicine addresses these inefficiencies by identifying the most effective interventions up front.

Healthier patients naturally contribute to economic productivity as well. When chronic conditions like diabetes, heart disease, and depression are managed effectively through personalized care, patients miss fewer workdays and maintain higher performance levels. One study estimated that improved health outcomes driven by precision medicine could increase quality of life and associated productivity for six major conditions by as much as $50 billion annually in the United States alone.[18] Additionally, reducing caregiving burdens allows families to reallocate time and resources, further contributing to economic growth.

Precision medicine also drives innovation and investment in healthcare technology. The demand for advanced genomic testing, wearable devices, and AI-driven solutions has created a booming sector that attracts significant private and public funding. Startups and established companies alike are investing in developing tools and platforms to support precision care, fostering job creation and economic development. It is estimated that $11 billion was invested in AI healthcare companies alone in 2024.[19]

Barriers to Realizing Cost Savings

Unfortunately, equitable access to precision medicine remains a challenge, particularly for underserved populations. Genomic testing, wearable devices, and advanced diagnostics are often cost-prohibitive for uninsured or underinsured patients. Without efforts to subsidize these technologies or include them in public healthcare programs, disparities in access could widen, limiting the reach and benefits of precision medicine to wealthier demographics.

At the same time, privacy concerns and data security regulations can slow the integration of precision medicine tools, as healthcare systems must navigate complex compliance requirements. Addressing these issues requires not only technical solutions but also public trust in the safety and ethical use of personal health data.

To fully realize the cost-saving potential of precision medicine, comprehensive changes are essential. This includes investments in infrastructure, equitable access initiatives, and policy reforms that support innovation while ensuring that precision medicine benefits all populations, not just a few. These challenges, though significant, are resolvable with coordinated efforts across healthcare, industry, and government sectors.

Technological advancements are making precision medicine more scalable and accessible every day, paving the way for its widespread adoption. The cost of genomic sequencing has plummeted, with prices expected to drop even further. Similarly, AI integration is streamlining data analysis and decision-making, reducing the time and expertise needed to interpret complex health information. These developments are enabling healthcare systems to implement precision medicine at scale, even in resource-limited settings.

Globally, precision medicine holds promise for addressing critical healthcare challenges. In countries with limited resources, scalable technologies like AI-powered diagnostic tools and wearable devices can help bridge gaps in care. For example, AI algorithms analyzing simple blood tests or smartphone images can detect diseases such as anemia or skin cancer early, reducing the need for costly, specialized interventions later. Wearable devices can support remote monitoring for chronic diseases, improving outcomes while minimizing strain on healthcare infrastructure.

To fully realize the potential of precision medicine, continued fundamental reforms and investment are essential. Policies supporting value-based care, equitable access, and integration of emerging technologies will be crucial in overcoming barriers. By embracing these advancements, healthcare systems worldwide can transition to a more efficient, proactive, and personalized approach, meeting the needs of diverse populations while addressing global healthcare challenges with innovative solutions.

The Universal Benefits of Precision Medicine

Precision medicine heralds a groundbreaking shift in how healthcare is delivered, offering a universal appeal that transcends traditional boundaries. Unlike the one-size-fits-all approach, precision medicine leverages advanced technologies to tailor treatments to an individual's unique genetic, environmental, and lifestyle factors. This innovative approach addresses three critical challenges in healthcare: effectiveness, efficiency, and equity.

By prioritizing treatments that are optimized for each patient, precision medicine enhances effectiveness, improving outcomes while reducing unnecessary interventions. Its focus on early detection and prevention minimizes healthcare inefficiencies, cutting costs and streamlining care delivery. Importantly, precision medicine also holds the potential to advance equity by making cutting-edge diagnostics and treatments more accessible, narrowing health disparities for underserved populations.

Personalized and Effective Treatment

By eliminating the guesswork traditionally associated with treatment decisions, precision medicine is redefining patient care. Through tailored therapies, precision medicine ensures that each patient receives care optimized for their unique biology, leading to faster recovery times, improved survival rates, and an enhanced overall quality of life.

Yet the benefits of precision medicine extend beyond clinical outcomes. By improving the predictability of treatments and empowering patients with data, it fosters a more proactive, engaged approach to health management. Using a wearable to track recovery metrics, for example, provides reassurance and enables providers to monitor progress remotely.

Precision medicine also transforms patient care by ensuring that treatments are not just effective but personalized and patient-centered. This approach gives individuals greater control over their health, creating a healthcare experience that is truly life-changing. For example, genetic testing for BRCA1 and BRCA2 gene mutations can reveal a heightened risk for breast or ovarian cancer, prompting patients to adopt proactive measures like regular screenings or preventative surgeries.

NOTE   The Breast Cancer 1 (BRCA1) and Breast Cancer 2 (BRCA2) genes are human genes that produce proteins responsible for repairing damaged DNA. They play a critical role in maintaining the stability of a cell's genetic material.

Similarly, Big Data analytics can integrate information from genetic tests, electronic health records, and population health studies to identify risk patterns for individuals and predict potential health issues with remarkable accuracy.

In the same way, preventative care significantly reduces the burden of chronic diseases on both patients and the healthcare system. Early detection allows for less invasive treatments, improved quality of life, and reduced hospitalizations. From a system perspective, it lowers healthcare costs by minimizing the need for expensive late-stage interventions.

Precision medicine integrates data, insights, and advanced analytics to create a powerful preventative care framework. By identifying risks early and promoting healthier lifestyles, it improves individual outcomes while also easing the strain on healthcare systems.

Cost Savings for Patients and Healthcare Systems

One of the most impactful aspects of precision medicine is the use of personalized diagnostics. As we have seen, genomic testing can identify specific genetic mutations driving diseases like cancer, allowing physicians to prescribe therapies that directly target the underlying cause. Unlike traditional treatments, which may involve broad-spectrum drugs with limited success rates, targeted therapies are more effective, reducing the need for prolonged and costly interventions. Studies show that these approaches can improve overall treatment effectiveness by minimizing hospitalizations and mitigating severe side effects that require additional care.[20]

Healthier populations resulting from precision medicine naturally lead to broader societal economic benefits. Early detection and personalized care help individuals manage chronic conditions effectively, reducing absenteeism and maintaining workplace productivity. For instance, diabetic patients using wearable glucose monitors can better control their condition, avoiding emergency hospital visits and time away from work.

Reduced caregiver burdens are another critical advantage. Chronic diseases often require extensive caregiving, leading to lost income and emotional strain for families. Precision medicine's focus on prevention and effective treatment minimizes these demands, enabling caregivers to redirect their energy toward personal or professional pursuits.

Precision medicine also accelerates advancements in pharmaceuticals and treatment methodologies. Targeted therapies allow drug developers to design treatments tailored to specific genetic mutations. This approach has led to breakthroughs like immunotherapies for cancer and gene therapies for rare diseases, offering patients more effective options with fewer side effects.

One example of transformative results comes from the Mayo Clinic, a leader in integrating precision medicine.[21] Through its Center for Individualized Medicine, Mayo uses genomic sequencing to guide cancer treatments, significantly improving effectiveness. The clinic also employs AI to analyze patient data, providing actionable insights for chronic disease management. This integration has streamlined workflows, reduced costs, and enhanced patient outcomes, setting a benchmark for healthcare innovation.

Beyond individual organizations, precision medicine fosters systemic innovation. It encourages collaboration between healthcare providers, tech companies, and pharmaceutical firms, driving advancements that benefit entire populations. By making care more efficient and effective, precision medicine not only transforms individual treatment but also enhances the broader healthcare system, ensuring better outcomes, reduced costs, and continuous innovation for the future of medicine.

Improving Access

Gratifyingly, precision medicine holds the potential to bridge gaps in healthcare by making advanced, personalized care accessible to underserved populations. Combined with supportive policies and targeted initiatives, precision medicine can help address challenges such as limited access, inadequate infrastructure, and financial constraints and reduce health disparities.

Key to this effort is the integration of real-time data gathering and telehealth platforms, which allow patients in remote locations to monitor their health and connect with providers without the need for frequent in-person visits. Mobile genomic testing units can bring advanced diagnostics to underserved areas, identifying risks and enabling early interventions managed by telehealth care managers.

Equitable access to precision medicine requires robust policies and funding. Government and private sector investments can subsidize genomic testing, expand telehealth infrastructure, and train healthcare workers to deliver precision care. Programs like the All of Us Research Program in the United States highlight the importance of inclusivity, as it aims to build a diverse database of health information to ensure that advancements in precision medicine benefit people of all backgrounds.[22]

Another example of a public health initiative addressing disparities is the implementation of AI-driven remote care systems in rural India.[23] These systems analyze patient data to detect conditions like diabetes and hypertension early, enabling timely treatment even in areas with limited healthcare access. Coupled with community health worker training, the initiative demonstrates how precision medicine can empower underserved communities.

By addressing disparities, precision medicine offers a pathway to more equitable healthcare. With targeted investments and inclusive policies, it can transform care for underserved populations, ensuring that personalized medicine benefits everyone, not just those in well-resourced areas. This approach not only improves individual outcomes but also strengthens public health systems, creating more sustainable population health.

Indeed, precision medicine is making waves worldwide, with national programs and international collaborations driving innovation and improving healthcare outcomes. In addition to the United States, countries like China, the UK, Japan, and Australia are leading the charge, showcasing the transformative potential of personalized medicine on a global scale (see Figure 3-2).

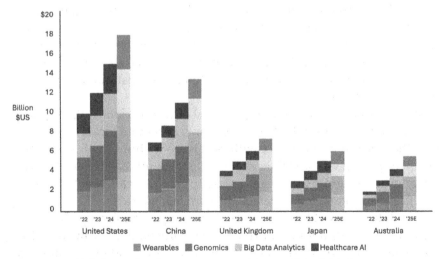

Figure 3-2: Global spending by country on precision medicine initiatives

Source: Adapted from [24]

China's Precision Medicine Initiative, launched in 2016, aims to integrate genomics, Big Data, and AI to improve diagnostics and treatment for its vast population.[25] By sequencing millions of genomes and analyzing extensive patient datasets, China is addressing chronic diseases and rare conditions with unprecedented precision, paving the way for scalable, personalized healthcare.

In the UK, the 100,000 Genomes Project set a global benchmark by sequencing the genomes of patients with rare diseases and cancer, positioning the National Health Service to leverage insights through successor programs like the Genomic Medicine Service.[26] This longitudinal initiative not only advances treatment for individual patients but also contributes to an expansive database, fostering global research collaborations. Similarly, Australia's Genomics Health Futures Mission focuses on integrating precision medicine into routine care, with a strong emphasis on improving equity and access for rural and Indigenous populations.[27]

International partnerships are accelerating these advancements, too. Collaborative research in genomics and AI between institutions across continents is uncovering new ways to predict, diagnose, and treat conditions. For example, global initiatives like the COVID-19 Host Genetics Initiative demonstrated how precision medicine can combat infectious diseases by identifying genetic factors influencing disease severity, enabling more effective interventions.[28]

By fostering global innovation and collaboration, precision medicine is not just transforming national healthcare systems—it's shaping a healthier, more connected world. These efforts highlight its potential to tackle widespread health challenges, making personalized care a cornerstone of global health equity.

Shared Future of Better Health

Precision medicine is a dynamic approach with the power to revolutionize healthcare on multiple levels. At the personal level, it tailors treatments to individual needs while streamlining care, cutting costs, and driving innovation in diagnostics and treatment methodologies for healthcare systems. On a global scale, it fosters international collaboration, bringing cutting-edge solutions to widespread challenges like chronic illness, infectious diseases, and health disparities.

As precision medicine transforms healthcare, however, its impact is perhaps most profound on those struggling to find answers in a system built on generalized approaches. The promise of wearable devices, as we saw with Michael and Jennifer's weight loss journey in Chapter 1, is undeniable, but for many patients, a personal breakthrough comes from pharmacogenomics—aligning medications with an individual's genetic makeup. This targeted approach has the potential to replace frustration with confidence, offering a clear path to treatment that works right the first time.

For Emily, this promise was still an unknown. A driven young woman, she had mastered the art of appearing in control—in both her professional and personal life. Yet beneath the surface, an exhausting battle with anxiety and depression had left her disillusioned. Like so many others, she had resigned herself to coping rather than thriving, until an unexpected conversation opened the door to a new possibility. Emily's journey, which we join in the next chapter, introduces her to medication optimization through genetics, a field that redefines not only her treatment but also her personal science of hope.

Notes

1. Colby, Brandon, "Whole Genome Sequencing Cost," January 17 (2025), https://sequencing.com/education-center/whole-genome-sequencing/whole-genome-sequencing-cost.

2. Jahangiri, Sonia, Abdollahi, Masoud, Rashedi, Ehsan, and Azadeh-Fard, Nasibeh, "A Machine Learning Model to Predict Heart Failure Readmission: Toward Optimal Feature Set," *Frontiers in Artificial Intelligence* 7 (2024).

3. Fallatah, Deema, and Adekola, Hafeez, "Digital Epidemiology: Harnessing Big Data for Early Detection and Monitoring of Viral Outbreaks," *Infection Prevention in Practice* 6, no. 3 (2024): 100382.

4. Davis, Nicole, "More Breast Cancer Cases Found When AI Used in Screenings, Study Finds," January 7 (2025), https://www.theguardian.com/society/2025/jan/07/more-breast-cancer-cases-found-when-ai-used-in-screenings-study-finds.

5. Majumder, Anusree, Sen, Debra, "Artificial Intelligence in Cancer Diagnostics and Therapy: Current Perspectives," *Indian Journal of Cancer* 58, no. 4 (2021): 481–492.

6. Huang, Kexin, Chandak, Payal, Wang, Qianwen, Havaldar, Shreyas, Vaid, Akhil, Leskovec, Jure, Nadkarni, Girish, Glicksberg, Benjamin,

Gehlenborg, Nils, and Zitnik, Marinka, "A Foundation Model for Clinician-Centered Drug Repurposing," *Nature Medicine* 30 (2024): 3601–3613.

7. Centers for Medicare & Medicaid Services, "National Health Expenditures 2022 Highlights," December 13 (2023), `https://www.cms.gov/newsroom/fact-sheets/national-health-expenditures-2022-highlights`.

8. Brown and Bussell, "Medication Adherence: WHO Cares?"

9. Sultana, Janet, Cutroneo, Paola, and Trifirò, Gianluca, "Clinical and Economic Burden of Adverse Drug Reactions," *Journal of Pharmacology & Pharmacotherapeutics* no. 4 (Suppl 1), December (2013): S73–S77.

10. Jarvis, Joseph, Peter, Arul, Keogh, Murray, Baldasare, Vince, Beanland, Gina, Wilkerson, Zachary, Kradel, Steven, and Shaman, Jeffrey, "Real-World Impact of a Pharmacogenomics-Enriched Comprehensive Medication Management Program," *Journal of Personalized Medicine* 12, no. 3 (2022): 421.

11. Hill, Harry, Roadevin, Cristina, Duffy, Stephen, Mandrik, Olena, and Brentnall, Adam, "Cost-Effectiveness of AI for Risk-Stratified Breast Cancer Screening," *JAMA Network Open* 7 no. 4 (2024): e2431715.

    Kacew, Alec, Strohbehn, Garth, Saulsberry, Loren, Laiteerapong, Neda, Cipriani, Nicole, Kather, Jakob, and Pearson, Alexander, "Artificial Intelligence Can Cut Costs While Maintaining Accuracy in Colorectal Cancer Genotyping," *Frontiers in Oncology* 11, June 8 (2021): 630953.

12. Chan, Bryan, and Hughes, Brett, "Targeted Therapy for Non-Small Cell Lung Cancer: Current Standards and the Promise of the Future," *Translational Lung Cancer Research* 4, no. 1 (2015): 36–54

13. Chenchula, Santenna, Atal, Shubham, and Uppugunduri, Chakradhara, "A Review of Real-World Evidence on Preemptive Pharmacogenomic Testing for Preventing Adverse Drug Reactions: A Reality for Future Health Care," *The Pharmacogenomics Journal* 24, no. 9 (2024).

14. Hannah, Katia, Nemlekar, Poorva, Green, Courtney, and Norman, Gregory, "Reduction in Diabetes-Related Hospitalizations and Medical Costs After Dexcom G6 Continuous Glucose Monitor Initiation in People with Type 2 Diabetes Using Intensive Insulin Therapy," *Advances in Therapy* 41, no. 6 (2024): 2299–2306.

15. Williams, Rachel, "Prostate cancer and AI: The Exciting Advances That Could Transform Treatment," January 24 (2025), `https://www.theguardian.com/research-to-reality/2025/jan/24/prostate-cancer-and-ai-the-exciting-advances-that-could-transform-treatment`.

16. Kasztura, Miriam, Richard, Aude, Bempong, Nefti-Eboni, Loncar, Dejan, and Flahault, Antoine, "Cost-Effectiveness of Precision Medicine: A Scoping Review," *International Journal of Public Health* 64, no. 9 (2019): 1261–1271.

17. Evans, William, "Dare to Think Rare: Diagnostic Delay and Rare Diseases," *British Journal of General Practice* 68, no. 670 (2018): 224–225.

    Kingsmore, Stephen, Cakici, Julie, Clark, Michelle, Gaughran, Mary, Feddock, Michele, Batalov, Sergey, Bainbridge, Matthew, Carroll, Jeanne, Caylor, Sara, Clarke, Christina, Ding, Yan, Ellsworth, Katarzyna, Farnaes, Lauge, Hildreth, Amber, Hobbs, Charlotte, James, Kiely, Kint, Cyrielle, Lenberg, Jerica, Nahas, Shareef, Prince, Lance, Reyes, Iris, Salz, Lisa, Sanford, Erica, Schols, Peter, Sweeney, Nathaly, Tokita, Mari, Veeraraghavan, Narayanan, Watkins, Kelly, Wigby, Kristen, Wong, Terence, Chowdhury, Shimul, Wright, Meredith, and Dimmock, David, "A Randomized, Controlled Trial of the Analytic and Diagnostic Performance of Singleton and Trio, Rapid Genome and Exome Sequencing in Ill Infants," *American Journal of Human Genetics* 105, no. 4 (2019): 719–733.

18. Dzau, Victor, Ginsburg, Geoffrey, Van Nuys, Karen, Agus, David, and Goldman, Dana, "Aligning Incentives to Fulfil the Promise of Personalized Medicine," *Lancet* 385, no. 9982 (2015): 2118–2119.

19. Landi, Heather, "1 in 4 Dollars Invested in Healthcare Going Toward Companies Using AI, Some Applying It to Precision Medicine," *Fierce Healthcare*, June 12 (2024), `https://www.fiercehealthcare.com/ai-and-machine-learning/1-4-dollars-invested-healthcare-going-toward-companies-using-ai-some`.

20. Li, Wenqian, Guo, Hanfei, Li, Lingyu, and Cui, Jiuwei, "Comprehensive Comparison Between Adjuvant Targeted Therapy and Chemotherapy for EGFR-Mutant NSCLC Patients: A Cost-Effectiveness Analysis," *Frontiers in Oncology* 11 (2021): 619376.

21. Curry, Timothy, and Lazaridis, Konstantinos, "Clinomics," January 24 (2025), `https://www.mayo.edu/research/centers-programs/center-individualized-medicine/research/pillars-programs/clinomics`.

22. All of Us Research Program, "Program Overview," November 22 (2024), https://allofus.nih.gov/about/program-overview.

23. Kerketta, Ajit, and Balasundaram, Sathiyaseelan, "Leveraging AI Tools to Bridge the Healthcare Gap in Rural Areas in India," *medRxiv,* preprint August 1 (2024), https://www.medrxiv.org/content/10 .1101/2024.07.30.24311228v1.full.

24. Mikulic, Matej, "Global Spending on Precision Medicine Treatments from 2022 to 2027," January 6 (2025), https://www.statista .com/statistics/1420946/spending-on-precision-medicine- treatments-globally.

Fortune Business Insights, "Precision Medicine Market Size, Share & Industry Analysis," January 6 (2025), https://www.fortune businessinsights.com/precision-medicine-market-110463.

Grand View Research, "Artificial Intelligence (AI) in Healthcare Market Size, Share & Trends Analysis," January 24 (2025), https:// www.grandviewresearch.com/industry-analysis/artificial- intelligence-ai-healthcare-market.

Tohoku University, "Tohoku Medical Megabank Organization," https://www.megabank.tohoku.ac.jp/english.

Australian Genomics, "Australian Genomics," https://www .australiangenomics.org.au.

25. Cyranoski, David, "China Embraces Precision Medicine on a Massive Scale," *Nature* 529 (2016): 9–10.

26. Genomics England, "100,000 Genomes Project," January 25 (2025), https://www.genomicsengland.co.uk/initiatives/100000- genomes-project.

NHS England, "NHS Genomic Medicine Service," January 25 (2025), https://www.england.nhs.uk/genomics/nhs-genomic- med-service.

27. Australian Government Department of Health and Aged Care, "Genomics Health Futures Mission," September 13 (2024), https:// www.health.gov.au/our-work/mrff-genomics-health-futures- mission.

28. COVID-19 Host Genetics Initiative, "COVID-19 hg," January 26 (2025), https://www.covid19hg.org.

Emily's Unrest

Emily darted through the lobby, heels clicking in rhythm as she balanced a coffee in one hand and her leather bag in the other. Her navy suit traced her frame closely, a reflection of her attention to detail. As she reached the elevator before it closed, she exhaled secretly, smoothing a stray lock of hair before stepping aboard.

By the time she arrived in the office, Emily's energy was infectious. "Morning, everyone!" she called out, her voice bright and alerting. She glanced at her assistant. "Jessica, the client deck is ready, right?"

"Of course," Jessica replied, her smile evident in her tone.

A Life of Contrasts

Emily's ability to juggle priorities was legendary, her colleagues often marveling at how she could transition from brainstorming a bold campaign tagline to negotiating contracts without missing a beat. "Creativity is just problem-solving with flair," she often quipped, her resourcefulness as much a part of her as her cutting-edge ideas. Leaders saw her as a vibrant personality, a presence who combined visionary thinking with a knack for galvanizing her team. For Emily, the fast pace wasn't chaos—it was home.

Emily's calendar was as full socially as it was professionally. Sunday dinners with her mom and her mom's second husband were a tradition, filled with her mother's probing questions about her love life and her stepdad's unsolicited career advice. "You're doing great, Emily, but remember to focus on the next big move," he'd remind her between bites. Emily would smile and nod, masking the weariness that came with these well-meaning but constant inquiries.

Her friends provided a welcome balance: easy laughter over avocado toast at the hotel brunch, giggly banter during game nights in her apartment, or relaxed evenings at house-hangs with friends. "You're so . . . adorable," her friend Rachel chided one evening, watching Emily charm a potential collaborator with her wit.

Yet beneath her polished exterior, Emily sometimes wondered if she was simply playing a role. The cheerful daughter, the reliable friend, the serious professional—each felt somehow rehearsed. Late at night, she'd replay conversations in her mind, questioning if they saw the real her or just the performance she'd given.

The Bright Light

Emily rose, her name announced to applause as she accepted the "Innovator of the Year" award. Her latest campaign, a groundbreaking blend of storytelling and technology, had not only exceeded revenue goals but also set a new standard in the industry. Smiling as she stood, she felt a swell of pride as her team cheered from their table.

"Congratulations again, Emily. You earned it," Michael, the company president, said afterward, toasting during the celebratory dinner. Emily nodded, her practiced smile in place. "Thank you, it really was a team effort," she replied, deflecting the spotlight with relieving grace.

But as the evening quieted and she returned home, the nagging voice crept in. *Was it luck? Could anyone else have done it better?* These thoughts, fleeting yet persistent, unsettled her. Outwardly, she was the picture of confidence and competence, yet internally, she wrestled with the fear of being exposed as less brilliant than others believed. Her accomplishments painted the image of a flawless future, but cracks in her self-assurance hinted at an inner battle still unfolding.

Each morning, Emily's day began with a carefully choreographed routine. The soft sunrise filtered through her apartment as she arranged her coffee cup, notebook, and flowers on the granite countertop. A quick photo for Instagram captured the serene moment: *Morning vibes,* she captioned, the likes coming immediately.

But behind the scenes, the picture was less perfect. Her alarm had been snoozed twice—or was it three times?—leaving her scrambling to apply makeup to her tired eyes. The coffee she photographed began to cool as she stared at her phone, apprehensive about the day's demands.

Inside, Emily felt the weight of anxiety and depression—a battle she hid behind her forced exterior. Few knew about the restless nights spent worrying over every decision or the moments of quiet despair when success felt hollow. The confident woman the world saw was real, but so was the exhausted, uncertain Emily she kept hidden. Her curated mornings masked a deeper struggle, one rooted in a lifetime of high expectations and the fear of falling short.

Emily grew up in a modest suburban neighborhood, the child of two determined parents whose marriage had crumbled but whose dedication to her remained. Weekdays were spent with her mother, a nurse who worked long shifts, and weekends with her father, a contractor who often brought her along to job sites.

"Don't forget your science project, Em," her mother called one morning, rushing out the door with her scrubs barely buttoned. That evening, her father's gruff voice echoed in the garage. "If you're going to work on your poster, be sure and clean up after."

The back-and-forth between households left Emily feeling untethered. Yet in the chaos, she discovered a sanctuary: school. She threw herself into it, excelling in art, her favorite subject, and contributing to the yearbook staff. Her teachers praised her congeniality, and her artwork adorned her bedroom walls.

Still, the emotional strain lingered. She learned to mask her vulnerabilities, to turn inward, and to push harder to replace the fear. Although these experiences shaped her resilience and drive, they also planted a seed of self-doubt. *If I don't prove my worth, will anyone stay?* The question followed her into adulthood, subtly steering her every move.

After high school, stepping onto the college campus for the first time, Emily felt a rush of exhilaration. The sprawling quads and historic buildings symbolized the independence she'd craved for years. But as the excitement settled, reality set in. Tuition was steep, and her parents' contributions, although generous, were barely enough. Part-time jobs at the library and a local café filled the gap but left little room for anything else.

Just a bit longer, she'd tell herself during late-night reading, her dorm illuminated by the dim light of her laptop. Despite the challenges, she succeeded. Her professors noticed the way others related to her, and

Emily threw herself into campus activities, from football games and class events to the (more than) occasional party. She found joy in the camaraderie of her friends, their laughter a welcome reprieve from the grind.

Still, beneath her achievements, anxiety quietly took root. *What if I can't keep up?* she wondered, a question that gnawed at her during quiet moments. College taught Emily resilience and confirmed her potential, but it also set the stage for the perfectionism and self-doubt she carried into her future.

Getting Closer

Emily still remembered the thrill of unlocking the door to her first apartment: a modest one-bedroom with creaky floors and a view of a busy street. It wasn't much, but it was hers. She furnished it piece by piece, balancing her entry-level salary with rent and student loan payments. Each paycheck was stretched to its limit, yet the sense of accomplishment felt good.

Her first job as a marketing associate was equally exhilarating and daunting. She quickly learned to navigate office dynamics, carefully wording emails and mastering the art of being confident yet agreeable. "Great idea, Emily," her manager once said in a meeting, validating weeks of painstaking effort.

Yet her relentless drive for perfection took its toll. Late nights at the office left her drained, and she often second-guessed her work. *Am I doing enough?* The question lingered, feeding an undercurrent of doubt.

Through the hurdles, Emily refined her skills, evolving into the confident young woman admired by her peers. But beneath the exterior, the pressure to excel never fully eased.

Sitting by her apartment window, Emily cradled a cup of tea, watching the city lights shimmer against the evening sky. Her reflection in the glass stared back, a polished professional with a string of achievements that once seemed unattainable. She had the career, the apartment, the accolades. By all accounts, she'd arrived.

Why isn't it enough? The thought crept in uninvited, its weight familiar yet unwelcome. She replayed the milestones in her mind—graduating college, landing her first job, winning the latest accolade. Each victory had brought a fleeting sense of fulfillment, only to be replaced by an insistent need to reach the next rung.

Frustration bubbled beneath the surface. *You have everything you wanted. Why can't you just be happy?* She felt guilty for feeling this way, as though her anxiety and bouts of depression betrayed the hard work that had gotten her here.

As the tea cooled in her hands, Emily realized her struggle wasn't with what she had achieved but with what she hadn't addressed. Beneath the layers of success lay questions she could no longer ignore—questions about what she truly needed to feel whole.

It always began with a familiar flurry of thoughts, her mind racing as she sped toward the office. She mentally cycled through her to-do list, rehearsed responses for a meeting, and worried whether she'd locked the door. *Did I double-check the report? What if I missed something?* The questions felt like static, always there, never quiet.

At work, she projected calm efficiency. Her emails were meticulously worded, every slide in her presentations double-checked, and her calendar organized with meetings. Staying busy was her armor, a way to keep the unease at bay. "You're always so prepared," a colleague once said, admiration in their tone. Emily smiled, but her shoulders seemed to ache from the tension she carried.

Despite her outward confidence, the physical manifestations of her anxiety told a different story. Her breaths sometimes came shallow, her chest tight, her heart speeding at unexpected moments. Alone in her apartment at night, she replayed conversations in her mind, dissecting her tone, her words, her gestures.

Why can't I just relax? she wondered, frustrated by her inability to switch off. Her anxiety was a quiet, constant presence—like the hum of an old refrigerator in the background of her life—persistent and draining, yet so familiar it went unnoticed by others.

Sunlight streamed through the blinds, casting a warm radiance on her bedroom this Saturday morning, but Emily felt none of its warmth. She stared at the ceiling, her mind a blur of self-recrimination.

You have so much to do—why can't you just get up? The thought circled relentlessly, compounding her guilt. She knew the list: laundry, groceries, finishing the outline for Monday's presentation. But even thinking about it felt insurmountable.

The fatigue wasn't physical—it was something deeper, an ache that seemed to drain her of energy and focus. On good days, she could push through it, masking her sadness with determination. But on days like this, it left her questioning everything. *What's the point of all this?* she wondered, her chest pounding with the weight of her thoughts.

Emily's perfectionism only deepened her despair. She berated herself for not being productive, for letting things slide. To the outside world, she was a beacon of success, but within, the unrelenting vacuum amplified her depression, turning her smallest struggles into what felt like insurmountable failures. She turned away from the light, wishing, futilely, for the feeling to pass.

Coping

The conference room buzzed with anticipation as Emily stepped to the front, the slide deck behind her. She portrayed confidence, her voice steady and engaging as she walked her team through a new campaign strategy. "This approach aligns perfectly with our target audience's values," she concluded, earning nods of approval and even a few comments of praise.

"Fantastic work, Emily," her manager said as the meeting wrapped up. "You really nailed it." Emily smiled, her shoulders straight and her posture poised. "Thank you. It's a team effort," she replied, her tone relievingly gracious.

Minutes later, she closed the door to her office and sank into her chair, the façade slipping away. Her chest pounded, and tears pricked the corners of her eyes. *Why do I feel this way?* she thought, pressing her palms against her temples. She'd just delivered a presentation most would envy, yet she felt hollow, as though she were a fraud teetering on the edge of discovery.

Her fear of judgment kept her silent. Opening up seemed impossible; what would people think of her if they knew the truth? So, Emily bore her struggles alone, the exterior masking the storm within, each success intensifying the contradiction she carried.

Emily buried herself in her work, finding solace in the mix of deadlines and campaigns. Her calendar was packed, leaving little room for introspection. When the office lights dimmed, she often headed to the gym, pounding out her frustrations on the treadmill. The endorphins helped, but only for a moment. By the time she returned home, the weight of her emotions settled back in.

During a break with Rachel, Emily finally voiced her frustrations. "I'm doing everything they say I should," she said, stirring her latte absently. "I go to therapy, I exercise, I even tried that new meditation app. But nothing seems to . . . stick."

"Maybe it just takes time?" Rachel offered hesitantly.

Emily forced a smile. "Yeah, maybe," she said, though her thoughts told a different story. *Time for what? More weeks of feeling like this?*

Her therapy sessions felt like going in circles, rehashing problems without progress. The medications she'd tried left her groggy or on edge, making her wonder if the side effects were worth it. *What if this is as good as it gets?* she wondered late at night, the thought both terrifying and numbing. She coped well enough to appear functional, but deep down, she yearned for more than survival—she wanted to thrive.

A Friendly Routine

The familiar hum of hairdryers and soft chatter greeted Emily as she stepped into Jennifer's salon. "Hey, Emily! Right on time," Jennifer said with a bright smile, ushering her to the chair.

As Jennifer draped the cape around her, Emily sank into the seat, the tension in her shoulders easing slightly. "You're a lifesaver," Emily sighed. "I've had the week from hell."

Jennifer chuckled, expertly sectioning Emily's hair. "Isn't that every week for you? You're always juggling a million things."

"I could say the same about you," Emily replied, admiring how Jennifer seemed to manage her bustling salon, three kids, and a thriving social life with ease. "How do you do it?"

"Honestly? I don't sweat the small stuff," Jennifer said, her tone warm but matter-of-fact. "Some days dinner's just mac and cheese and that's okay. It's about balance."

Emily smiled, wishing she could adopt even a fraction of Jennifer's perspective. Their conversations always felt easy, a reprieve from the constant demands of her life. As Jennifer styled her hair, Emily found herself reflecting on her own struggles, wondering if she'd ever find that elusive balance. For now, the salon chair was her safe haven, a rare moment to simply breathe.

As Jennifer moved around her, expertly shaping her hair, Emily couldn't help but admire the ease with which Jennifer seemed to navigate life. Her stories of managing a busy salon, helping with her kids' school projects, and still squeezing in yoga sessions painted a picture of balance and fulfillment.

"You make it look so easy," Emily said, half-joking. "If I had half your energy, maybe I'd finally have it all figured out."

Jennifer laughed. "Oh, trust me, it's not always easy. Some days I'm just winging it. But I try to focus on what really matters—and let the rest go."

Emily nodded, but a pang of envy lingered. Jennifer's life seemed so full, yet grounded, a stark contrast to her own, which often felt like an endless sprint with no clear finish line.

Still, their conversations left her inspired. Jennifer's positivity and intentionality made Emily wonder if she, too, could find a way to slow down and prioritize the things that truly mattered. *Maybe it's time to make some changes*, she thought, though the prospect felt overwhelming. For now, she carried Jennifer's words with her, a small spark of hope that perhaps life didn't have to be quite so frantic.

Arriving home, the soft hum of the refrigerator filled Emily's apartment, the only sound breaking the silence. She sat on her couch, opening her laptop's beckoning screen, a half-written email waiting for her to finish. The words blurred together as she rubbed her temples, her mind unable to summon the focus needed to finish the message and hit "Send".

Around her, the evidence of her fatigue reappeared. A stack of unopened mail teetered on the edge of her kitchen counter, bills and promotions mingling in neglected disarray. Across the room, the laundry basket overflowed, its contents spilling onto the floor—a silent reminder of the tasks she kept putting off.

The ticking of the wall clock echoed faintly, marking time she felt slipping away. She glanced at it, a pang of guilt tightening her chest. *I should just get up and do it. It's not that hard*, she told herself, but her body refused to move.

Her procrastination only fueled her anxiety. Each unfinished task became a symbol of failure, adding to the growing weight she carried. As the laptop screen dimmed, Emily sank further into the couch, overwhelmed by the sheer number of small things she couldn't bring herself to tackle. The apartment felt suffocating, a reflection of the clutter in her mind.

Ending the day, Emily lay in bed, staring at the faint outline of the ceiling fan as it spun lazily above her. The city outside her window had quieted hours ago, but her mind refused to follow suit. Thoughts darted in every direction, colliding and multiplying. *Did I forget to email the client back? What if they think I'm unprofessional?* The tasks of the day morphed into worries about her future. *What if I've peaked? What if this is all there is?*

Sleep came, but fitfully. Emily shifted onto her side, slapping her pillow in frustration. The clock on her nightstand glared 2:47 a.m. *Why can't I just shut it off?* The harder she tried to relax, the more elusive sleep was. Her heart raced, as though her body were bracing for a disaster that existed only in her head.

Beneath the frustration lay a deeper fear: *What if I never feel at peace?* The thought was sobering, heavy in the stillness of the night. Yet, as the minutes ticked by, another thought emerged, tentative but determined. *This can't keep going. Something has to change.*

For the first time in weeks, she allowed herself a sliver of hope. Maybe it was time to explore new ways to address her mental health—beyond coping, beyond surviving. With that thought, she closed her eyes, hoping sleep might finally come.

Breaking Point

Emily sat at her desk, the glimmer of her computer screen casting shadows across her face. The cursor blinked impatiently in a document titled "Campaign Proposal – Final Draft." The deadline was just hours away, but her mind felt like a tangled web of unfinished thoughts. She typed a sentence, deleted it, and sighed.

The struggle with the page stole more time than she thought. A notification pinged, jolting her. She opened it to see a message from her manager: "Can we review the draft at 3:00 p.m.?" Her stomach clenched. She glanced at the clock—barely two hours to pull everything together.

Her assistant peeked in hesitantly. "Hey, Emily, the client meeting started five minutes ago. Did you want me to—?"

Emily froze. *The meeting. How could I forget?* "No, no, I'm coming," she said quickly, grabbing her notebook and forcing a smile. "Thanks, Jessica."

Inside, panic churned. Forgotten meetings, rushed reports, and rising frustration from her team had become unsettlingly frequent. *What's happening to me?* she wondered, guilt tightening her chest.

As the meeting dragged on, Emily's thoughts spiraled. *They're going to notice. They'll realize I'm slipping. What if they think I can't handle this?*

The fear of being exposed as incompetent only made things worse. By the time she returned to her desk, the vicious cycle had consumed her: overthinking led to underperforming, which only deepened her anxiety. The once-confident young woman was crumbling under the weight of her own expectations.

The conference room felt uneasy as Emily flipped through a stack of slides. Her team watched quietly, sensing her agitation. "This isn't what we discussed," she said sharply, pointing at a bar chart. "We agreed on a different approach, didn't we?"

"Actually," Ben, her colleague, interjected cautiously, "we adjusted it based on the client's updated metrics from last week's call. Do you remember the email?"

Emily's jaw tightened. "I don't remember agreeing to that," she replied, her tone defensive. The room fell silent, the awkwardness palpable.

Ben nodded diplomatically. "No problem. Let's revisit it later," he offered, although his polite smile faltered as Emily quickly changed the subject.

Back at her desk, Emily replayed the interaction in her mind. *Did I sound unprepared? Am I losing my edge?* The thought made her stomach churn. She knew her reactions were off—too curt one moment, too withdrawn the next—but her mounting stress made it hard to correct.

Her team had always admired her intuition and precision, but lately, she'd noticed their hesitation to approach her. The once-collaborative dynamic now felt strained, a distance growing between Emily and the colleagues who had once been her strongest allies. She hated the idea that her struggles were affecting them, but she didn't know how to stop the spiral.

Arriving home, Emily stared at her phone, a text from Rachel blaring on the screen: "Game night at 7! Don't forget—it's your turn to bring snacks!" Her fingers hovered over the phone before she typed, "Can't make it tonight, swamped with work. Sorry!" She hit "Send", a pang of guilt settling in as she looked around her quiet apartment. The unopened mail, opened takeout box, and muted TV painted a different picture than "swamped."

Her phone buzzed again—this time her mother. Emily hesitated before answering. "Hi, Mom," she said, her tone curt and flat.

"Hi, sweetheart! We haven't seen you in weeks. How about dinner Sunday?"

Emily pinched the bridge of her nose. "I can't, Mom. Work's crazy right now."

Her mother's voice softened. "Emily, you need to take care of yourself. You're working too hard."

"I'm fine," Emily replied quickly. "Just busy. Gotta go—early meeting tomorrow."

As she ended the call, the guilt tightened in her chest. She knew her parents were worried, their invitations more frequent, their concern evident in their voices. But explaining how she felt seemed impossible. *How do I tell them I'm too tired to face anyone, even them?*

Her isolation grew with each canceled plan and brief phone call, leaving her lonelier and more ashamed for pulling away from the people who cared most about her.

Emily sat on her couch, staring blankly at the coffee table. It was cluttered with papers, an empty coffee mug, and the remnants of last night's takeout. The apartment, once a reflection of her polished nature, had become a silent testament to her inner turmoil. Dust gathered on her bookshelves, laundry spilled from the basket in the corner, and the faint scent of forgotten food lingered in the air.

She glanced at her laptop, its screen dark and untouched. A simple task—updating a single slide for tomorrow's presentation—felt impossible. *Just open it. Five minutes, that's all you need,* she told herself, but her body refused to move. The weight pressing down on her was as heavy as it was invisible.

Outside her apartment, Emily still tried to be the perfecting professional. She started meetings with a smile, exchanged pleasantries with colleagues, and delivered presentations with forced confidence. But at home, that persona crumbled. Here, she wasn't the accomplished professional but a woman paralyzed by her own mind, each day blurring into the next.

How did I get here? she wondered, her chest pounding with guilt and frustration. The contrast between her public success and private despair was a chasm she didn't know how to bridge, leaving her feeling trapped in a life that no longer felt like her own.

Darkness, My Old Friend

The email read: "URGENT: Missing Campaign Proposal." Emily's stomach dropped. She'd forgotten to submit the final draft, a key deliverable for a high-profile client. Her hands trembled as she hastily typed an apology, but it was too late—the damage was done.

Minutes later, her manager called her into the office. "Emily, this isn't like you," he began, his tone firm but concerned. "What's going on? Missing this deadline puts us in a tough spot."

Emily's mind raced as she tried to respond. "I—I've been overwhelmed," she stammered, avoiding his gaze. "It won't happen again."

Her manager sighed. "We all have a lot on our plates. I need you to be honest with me. Are you okay?"

She nodded quickly, forcing a weak smile. "I'm fine. Just a bad week."

But as she left the office, shame and helplessness overwhelmed her. *Am I fine?* she wondered. *Why can't I get it together?*

Sitting at her desk, Emily stared blankly at her screen, tears threatening to spill. She couldn't keep living like this—something had to change. For the first time, she began to consider that maybe she couldn't fix this alone. It was time to find help.

Emily's journey with mental health treatments began in her sophomore year of college. Overwhelmed by mounting academic pressure and social expectations, she made her way to the campus health center. The counselor listened patiently before suggesting she try medication for her anxiety. Desperate for relief, Emily agreed, clinging to the hope that this would bring the peace she craved.

The weeks that followed were a blur of adjustments. One medication left her too fatigued to focus in class; another caused her to gain weight, amplifying her insecurities. By her junior year, she'd cycled through half a dozen options, each requiring weeks of waiting, hoping the side effects would fade and the benefits would take hold.

"I feel like a science experiment," she vented to a friend during one particularly discouraging period. "It's like they're just guessing, trying to see what sticks."

Although some treatments dulled her symptoms, they often left her feeling emotionally flat, as though she were watching her life from a distance. The process was exhausting, and progress felt elusive.

By the time she graduated, Emily had resigned herself to coping rather than thriving, the trial-and-error process chipping away at her hope. Now, years later, the same frustrations lingered, leaving her questioning whether true relief was even possible.

Emily sat on the familiar couch in her therapist's office, staring at the soft patterns on the rug. "So, what's been bothering you this week?" her therapist prompted gently.

Emily hesitated, her mind racing to organize her thoughts into something coherent. *Where do I even start?* she wondered, frustration bubbling beneath the surface. Finally, she said, "I don't know . . . it's just everything. Work, life—it's all too much, but I'm fine. I'll figure it out."

Her therapist gave her a patient but probing look. "Emily, you say you're fine, but you're here for a reason. Let's try to dig into that."

Emily nodded but felt stuck, unable to bridge the gap between her thoughts and her words. She left the session feeling as though she'd failed, even in therapy.

Her perfectionism made it hard to fully engage. She worried about saying the "wrong" thing or not having clear answers, leaving her skimming the surface of her struggles instead of diving deep. Sessions often felt repetitive, focused on coping strategies she already knew but couldn't seem to implement. Therapy, although helpful at times, hadn't yet provided the breakthrough she desperately needed. Instead, it felt like another arena where she fell short of her own expectations.

Later that evening, the blue screen of Emily's laptop illuminated her tired face as she scrolled through a forum titled "When Nothing Works: Coping with Anxiety and Depression." Post after post mirrored her own frustrations—stories of trial and error with medications, therapy sessions that felt like dead ends, and the gnawing fear of never feeling better.

Why does this have to be so hard? she thought, scrolling past a comment about someone finding relief in meditation retreats, another touting an experimental treatment. She clicked on a thread discussing alternative therapies, her skepticism battling with a glimmer of curiosity.

She leaned back in her chair, rubbing her temples. *I've tried the "right" way—medications, therapy, self-help books—and I'm still stuck.* The thought stung, deepening her disillusionment with the traditional health system. It felt like an endless cycle of promises without results, a system that viewed her as a checklist rather than a person.

Yet, amid her frustration, a small voice whispered: *Maybe it's time to try something different.* Emily wasn't convinced, but the idea of continuing as she was felt unbearable. Closing her laptop, she resolved to keep looking, uncertain of what she might find but open, at last, to the possibility of something new.

As Emily lay in bed, her mind drifted back to the forums she'd scrolled through earlier. *Is there really something out there that could work for me?* she wondered. The idea felt fragile, like a flicker of light in a dark room. She longed for a solution that wouldn't just mask her symptoms but truly help her feel whole again. Yet, doubt lingered. *What if this is just how life is—always managing, never thriving?*

A Light

The next day, during her regular appointment with Jennifer, the conversation took an unexpected turn. Jennifer's salon was always a sanctuary for Emily, a rare place where the weight of her life felt just a little lighter. As Jennifer worked on her hair, chatting about her kids and their latest adventures, Emily felt a flicker of warmth.

"How do you juggle everything and still stay so . . . together?" Emily said.

Jennifer smiled as she sectioned Emily's hair. "Honestly? I don't have it all together. Some days are chaos, but I try to focus on the positives. And when it gets tough, I talk it out with people I trust."

Emily hesitated, staring at her reflection in the mirror. "I wish I could do that," she murmured.

Jennifer paused, her hands still for a moment. "Do what?"

"Talk about it. Everything," Emily said softly. "Lately, it just feels like . . . like I'm drowning. Work, life, everything—it's so much."

Jennifer met her eyes in the mirror, her expression gentle. "Emily, that sounds so hard. Sometimes, just saying it out loud can help."

Emily felt her chest loosen slightly. "Yeah," she admitted, her voice almost a whisper. "I guess it does."

Jennifer squeezed her shoulder lightly. "You don't have to figure it all out at once. But you don't have to go through it alone, either."

For the first time in weeks, Emily felt a small measure of relief. In this simple, unguarded moment, she realized how much she'd needed to share her feelings—even if only briefly.

As Jennifer worked on Emily's hair, the conversation turned more personal. "You know," Jennifer began, clipping a strand into place, "a couple of years ago, I hit a wall too. Juggling work, the kids, and everything else—it all caught up to me. I wasn't sleeping, my energy was gone, and my moods were all over the place."

Emily raised an eyebrow. "You? I can't picture that."

Jennifer chuckled. "Oh, trust me, it was rough. But things changed when I started *paying attention to me*. I got this watch that tracks everything—heart rate, sleep patterns, stress levels. It helped me understand my body better, like how my sleep was affecting my mood or when I needed to take a break."

Emily shifted in her seat, unsure. "I don't know . . . I've tried so many things. Nothing really works."

Jennifer placed a hand on Emily's shoulder. "I get that, but this isn't about fixing everything overnight. It's about understanding what you need. Maybe it's time to talk to someone who can look at your health in a new way—someone who'll really listen."

Although resistant, Emily couldn't deny a hint of curiosity. Jennifer's story made her wonder: could personalized care offer the solution she'd been searching for?

As Emily stepped out of the salon, the crisp air brushed her face, carrying with it a faint sense of hope. Jennifer's words lingered: "It's about understanding what you need." She wasn't sure what her next step would be, but for the first time in a long while, she felt the smallest possibility—a seed planted, waiting to grow.

The next morning, following a restless night, Emily hesitated, her finger hovering over the Call button on her phone. The number for Dr. Harper, a psychiatrist her colleague swore by, stared back at her from the screen. Anxiety twisted in her chest. What if this is just another dead end? *she thought, her mind flashing back to the countless appointments that had left her feeling like a puzzle no one could solve.*

But Jennifer's words echoed in her mind: "It's about understanding what you need." The memory of Jennifer's steady encouragement, paired with her own growing exhaustion, nudged her forward. With a deep breath, she tapped the button, her heart pounding as the receptionist answered.

"I'd like to schedule an appointment with Dr. Harper," Emily said, her voice steady despite the nerves tightening her throat.

After hanging up, she sat quietly for a moment, letting the decision sink in. A faint glimmer of hope broke through her doubt. *Maybe this time will be different,* she thought, though the skepticism lingered. As uncertain as she felt, making the call was a step forward—a small but significant act of choosing to fight for herself again.

Arriving later that week, Emily found that Dr. Harper's office was nothing like the sterile clinics she had visited before. Soft lighting, a cozy armchair, and a bookshelf filled with titles on mindfulness and behavioral health created a warm, inviting space. Dr. Harper herself, with her kind eyes and calm demeanor, greeted Emily with a hand that felt reassuring and warm.

"Tell me what brought you here today," Dr. Harper said gently, her voice steady but inviting.

Emily hesitated, her hands fidgeting in her lap. "I guess . . . I'm just tired. Not physically, but mentally. I feel like I'm failing, even when I'm not."

Dr. Harper nodded her expression attentive. "That sounds incredibly difficult. Can you tell me more about when this started?"

As Emily began recounting her story—college stress, the loop of medications, therapy that never quite worked—she felt a surprising sense of ease. Dr. Harper listened without interrupting, occasionally asking a clarifying question or jotting down a note.

For the first time in a long while, Emily felt truly heard. There was no rush to fix her, no dismissive suggestions or platitudes. Just a compassionate listener who seemed genuinely invested in understanding her. By the end of the session, an ember of hope had been rekindled. *Maybe I've finally found someone who can help*, Emily thought.

Toward the end of their session, Dr. Harper leaned forward slightly, her tone thoughtful. "Emily, I want to introduce you to something that might help us better understand how to treat your symptoms. Have you heard of pharmacogenomic testing?"

Emily shook her head. "Not really. What is it?"

"It's a way to analyze your DNA to see how your body processes medications," Dr. Harper explained. "Instead of guessing which treatments might work, we can use this information to personalize your care. It could help us avoid medications that might cause side effects or won't be effective for you."

Emily tilted her head, curiosity rising. "So . . . it's not just trial and error?"

"Exactly," Dr. Harper said with a small smile. "It's about using science to guide our decisions, making the process more precise."

Emily hesitated, her skepticism surfacing. *Another test, another promise*, she thought, though something about Dr. Harper's calm confidence made her consider it.

"Okay," Emily said finally, her voice cautious. "Let's do it."

As Dr. Harper explained the next steps, Emily felt a shift within her. She wasn't convinced this would work, but for the first time in years, she felt like she was taking a step forward with purpose. It was a small moment, but it marked the beginning of something new.

Pharmacogenomics: A Tailored Solution

Emily sat in Dr. Harper's office, a small kit in front of her on the desk. "That's it?" she asked, raising an eyebrow.

Dr. Harper smiled. "That's it. Just a simple cheek swab."

Emily picked up the swab, rolling it between her fingers. She had imagined something far more invasive or complicated—a blood draw, perhaps, or hours spent hooked up to some machine. But this? It seemed almost too easy. Following Dr. Harper's instructions, she rubbed the swab gently inside her cheek, then placed it into the labeled tube.

"There," Dr. Harper said reassuringly. "We'll send this off, and in about a week, we'll have a clearer picture of how your body responds to different medications."

As Emily watched the test kit being placed into the envelope, a feeling of empowerment stirred within her. For once, she wasn't just waiting for a generic solution or relying on guesswork—she was actively participating in her care.

Still, doubt lingered in her mind. *What if it doesn't help? What if this is just another thing that leads nowhere?* The weight of needing this to work pressed down on her, but curiosity pushed her forward. *Maybe this really could change things,* she thought, holding onto that small, fragile hope as she left the office.

The days following the test felt both endless and fleeting. Emily threw herself into her usual routine—meetings, emails, workouts—but the waiting was always there, humming in the background. She checked her phone obsessively, refreshing her email constantly. Each time, the absence of a notification brought both relief and frustration.

At work, her focus wavered. She caught herself zoning out during a meeting, her mind wandering to what the results might say. *What if they find something wrong?* she wondered, her stomach tightening. *Or worse, what if they don't find anything at all?*

One evening, unable to quiet her thoughts, Emily pulled out her journal. The pages were mostly blank—she'd always struggled to keep up the habit—but tonight, the act of writing felt necessary.

What am I hoping for? she scrawled across the top of the page. After a pause, she wrote: *Answers. Something to help me understand why I feel this way—and how to fix it.*

She tapped the pen against the paper, hesitating before adding: *What if it doesn't help? What if this is just another thing that goes nowhere?*

Setting the journal aside, she leaned back and stared at the ceiling. The wait felt heavier than she'd expected, but beneath the fear, a fragile thread of hope remained. *At least I'm doing something,* she reminded herself. *That has to count for something.*

Emily's phone buzzed on her desk, the subject line of the email catching her eye: "Your Test Results." Her heart skipped a beat as she stared at the screen, her fingers hovering over the notification. *This is it,* she thought, excitement and anxiety swirling in equal measure.

She took a deep breath before opening the email. The words blurred for a moment as she scanned the detailed summary. The report outlined her genetic markers and how they influenced her response to various medications. Specific recommendations followed, highlighting which treatments were likely to work best and which ones to avoid.

She read the analysis twice, the clear explanations making the once-mysterious process feel almost approachable. Her usual skepticism lingered at the edges of her thoughts. *It's just another plan,* she reminded herself. But this time felt different. This wasn't guesswork—it was tailored to her, based on her own biology.

Setting down the phone, Emily let out a breath she hadn't realized she was holding. A cautious optimism stirred within her. *Maybe this is what I've been waiting for,* she thought. The fear of disappointment still lingered, but for the first time in years, it felt like she was standing on the brink of something real—something that might finally bring her the clarity and relief she desperately needed.

Later, at the appointment with Dr. Harper, Emily sat across from her, the test report spread out between them. Dr. Harper's pen hovered over the paper as she began to explain. "Your results show a specific genetic variation in your DNA. This explains why medications like Lexapro or Zoloft didn't work as well for you—they're metabolized by this gene, and your body processes them too quickly to be effective."

Emily frowned, shifting in her seat. "So, does that mean there's something . . . wrong with me? Like I'm abnormal or something?"

Dr. Harper shook her head gently. "Not at all. Variations like this are incredibly common. In fact, it's just one of many factors that influence how we respond to medications. It's not about something being 'wrong'—it's about understanding how your body works so we can make better choices for your care."

Emily nodded slowly, her shoulders relaxing slightly. "So, it's not as rare or unusual as it feels?"

"Not at all," Dr. Harper reassured her. "It's just information—information we can use to help you feel better."

Emily leaned forward, her curiosity overcoming her skepticism. "How do you know this will work? I've tried so many things before."

Dr. Harper's tone remained steady. "It's not a guarantee, but this approach eliminates much of the guesswork. It's based on how your body processes medications, not on trial and error."

Emily felt a mix of relief and cautious hope. *Finally, something that makes sense,* she thought, though a part of her still braced for disappointment. As Dr. Harper outlined the next steps, Emily allowed herself to believe—just a little—that this time might be different.

The first day of Emily's new medication plan began with her sitting at her kitchen table, the small pill bottle in her hand. She read the label twice, as though the instructions might reveal a hidden secret, then swallowed the tablet with a sip of water. The action was small, but it felt monumental.

Over the next few days, Emily became hyperaware of every sensation in her body. A twinge of nausea after breakfast had her scanning the side effects list, and a faint headache by mid-afternoon made her wonder if she'd already made a mistake. She sat at her desk, fingers drumming nervously, her thoughts darting between hope and skepticism. *What if this is just like before? What if it doesn't work?*

That evening, as she brushed her teeth, she stared at her reflection in the bathroom mirror. The doubt in her eyes was familiar, but so was the small spark of determination beneath it. *It's too soon to know,* she reminded herself, her grip tightening on the counter. *Give it time.*

The days stretched on, and Emily began to notice subtle changes—a deeper sleep, a moment of focus at work that felt easier than usual. They were small victories, but they kept her moving forward, cautiously optimistic that this new plan might be the one to finally make a difference.

Emily tapped her pen against her notepad as she prepared for the morning meeting. A client had raised concerns about the latest campaign—a scenario that would usually leave her stomach in knots. But today, something felt different. Her thoughts weren't racing uncontrollably, and the tightness in her chest wasn't there. Instead, she felt . . . steady.

As the meeting began, Emily calmly addressed the client's concerns, her voice confident and measured. "I understand your feedback, and I think we can make adjustments without compromising the core message. Here's what I propose," she said, outlining a revised plan. When the client nodded in agreement, relief washed over her, and so did a sense of pride.

Back at her desk, she allowed herself a moment to reflect. *Was that . . . easier than usual?* she wondered. The meeting had gone well, but more importantly, she hadn't felt the usual wave of anxiety beforehand.

Over the following days, other small changes began to emerge. She woke up feeling rested rather than drained, completed tasks without the constant weight of self-doubt, and even caught herself smiling during a quiet moment at lunch.

It's too soon to call this a success, she reminded herself, her cautious nature keeping her grounded. But the shifts, however subtle, felt significant. For the first time in years, Emily began to believe that better days might be possible.

A Renewed Emily

One evening, Emily found herself at her kitchen counter, a recipe book propped open beside her. The ingredients for a dish she hadn't made in years—her favorite pasta bake—were scattered around. As she chopped herbs, the rhythmic motion felt surprisingly soothing, a small reminder of the joy she'd once found in cooking.

Later, with the scent of garlic and basil filling her apartment, she picked up her sketchpad, untouched for months. Her pencil glided over the page; the lines tentative at first but growing more confident with each stroke.

These moments felt like reconnecting with an old friend—herself. The improvements in her energy and focus gave her a glimmer of cautious excitement. *Maybe this is the start of something better,* she thought, allowing herself a small, hopeful smile. For the first time in years, the future didn't feel so heavy. Instead, it felt . . . possible.

That weekend, Emily sat on her balcony, the soft hum of the city below her, a cup of tea in hand. For the first time in years, her thoughts weren't racing, nor was her chest heavy with the weight of anxiety. Instead, her mind felt clear, her emotions steady.

Is this what normal feels like? she wondered, taking a sip of tea. The question lingered, bittersweet. She had been so consumed by anxiety and depression that they had become a part of her identity. Without them, she felt lighter but almost unfamiliar to herself, as though rediscovering a long-lost friend.

With the stability came a renewed confidence in her capabilities. At work, she found herself contributing more ideas, feeling less hesitant to speak up. Even her personal life felt richer—cooking, drawing, and calling her parents felt natural again, not like insurmountable tasks.

I'm still me, she realized, a small smile forming. *But now, I'm more.* The changes weren't monumental or immediate, but they were enough to remind Emily of the person she had always been beneath the struggle: resilient, capable, and whole. For the first time, she felt a cautious but genuine belief that her best days might still lie ahead.

One Sunday, Emily sat across from Rachel at the hotel brunch, the hum of chatter around them. It was the first time they'd done this together in months, and Emily felt the difference immediately. Her energy wasn't forced, her smile genuine.

"I missed this," Rachel said, her voice soft but sincere. "I missed you."

Emily looked down, guilt building in her chest. "I'm sorry," she said quietly. "I know I pulled away." She paused, searching for the right words. "I was . . . struggling more than I wanted to admit. But I'm trying now. I'm better."

Rachel reached across the table, squeezing Emily's hand. "I'm just glad you're here now."

That conversation was one of many. Emily found herself answering her parents' calls with warmth, rather than obligation, and accepting their invitations to dinner without hesitation. At work, she apologized to a colleague for her defensiveness in past meetings, working to rebuild trust.

As Emily mended these connections, she felt a growing sense of gratitude. The people in her life hadn't given up on her, even when she had distanced herself. Each small step toward reconnection reminded her of the strength in vulnerability and the importance of leaning on others.

The laughter of her team echoed around the office as Emily raised her glass of flavored water, toasting their latest campaign success. The warmth of the moment enveloped her, the genuine camaraderie of her colleagues filling the space. She found herself smiling—not out of obligation, but because she truly felt the joy of the celebration.

Later that evening, Emily curled up on her couch with a book, the soft glow of her reading lamp casting a cozy light. The aroma of her decaf tea wafted from the cup, and the rhythmic ticking of the clock blended with the gentle hum of the city outside her window. For the first time in years, she wasn't replaying conversations or planning tomorrow's to-do list. She was simply present, absorbed in the story unfolding on the page.

This is what I've been missing, she thought, the realization bringing a sense of quiet contentment. Life wasn't perfect, but it felt manageable, even beautiful in its simplicity.

Each moment she embraced reinforced a newfound hope—a belief that the future held not just challenges, but possibilities. Emily was no longer merely surviving; she was beginning to truly live.

Sitting on her balcony as the sun dipped below the horizon, Emily allowed herself a rare moment of reflection. The weight she had carried for so long no longer defined her days. The memories of sleepless nights, racing thoughts, and the constant ache of self-doubt felt distant, although not forgotten.

I've come so far, she thought, her lips curving into a small smile. The journey hadn't been easy. There were setbacks, doubts, and days when she wanted to give up. But through it all, she had found a resilience she had forgotten.

Emily knew her health journey wasn't over. There would still be challenges, moments of doubt, and days that felt heavier than others. But now, she had tools, support, and—most importantly—hope.

As she gazed at the fading light, her mind wasn't consumed with worry about tomorrow. Instead, she felt a cautious but genuine excitement for what lay ahead. *I'm ready for this,* she thought, taking a deep breath.

For the first time in years, Emily saw her future not as a burden to be endured but as a path to be explored. She felt ready to embrace it, one step at a time.

Genetics and Medication Management

For decades, prescribing medications has largely been an exercise in trial and error. Physicians follow established guidelines, selecting drugs based on population-level effectiveness, yet patients often experience vastly different outcomes. Some respond well, achieving the intended relief or disease control, whereas others suffer from side effects, treatment failure, or dangerous adverse reactions. This variability isn't due to chance—it's rooted in the fundamental differences in how individuals metabolize and respond to medications. Until recently, the tools to predict these differences were limited, leaving both doctors and patients navigating a frustrating, inefficient system.

As we saw in the previous chapter with Emily and her anxiety and depression challenges, precision medicine, particularly in the realm of genetics and medication management, changes this paradigm. By analyzing a patient's genetic makeup, we can predict how their body will process and respond to specific drugs, before they ever take the first dose. This approach, known as *pharmacogenomics*, enables physicians to select the right medication at the right dose for each individual, significantly reducing the risks of adverse reactions, nonresponse, and unnecessary suffering. As this chapter explores, integrating genetics into medication management is not just an improvement to current prescribing practices—it's a fundamental shift toward safer, more effective, and cost-efficient care.

Medication Failure in Traditional Care

Medication failure occurs when prescribed treatments fail to achieve the desired therapeutic outcome due to factors such as improper drug selection, incorrect dosing, nonadherence, or unanticipated side effects. For patients, this often leads to prolonged illness, preventable complications, and unnecessary suffering. For the healthcare system, medication failure represents a critical inefficiency, contributing to avoidable hospitalizations, repeated doctor visits, and increased reliance on emergency care.

Globally, the financial toll of medication failure is staggering, with estimates in the hundreds of billions of dollars annually. In the United States alone, studies suggest that nonoptimized medication regimens account for more than $600 billion in avoidable costs each year, representing a significant drain on an already strained system.[1] Beyond economics, these failures deepen inequities in healthcare, disproportionately affecting those with limited access to personalized care or resources to address the consequences.

Understanding the statistics behind medication failure is essential for recognizing the gaps in traditional care and the urgent need for precision medicine.

Understanding the statistics behind medication failure is essential for recognizing the gaps in traditional care and the urgent need for precision medicine. By tailoring treatments to individual patients based on their genetic, environmental, and lifestyle factors, precision medicine offers a pathway to reducing inefficiencies, improving outcomes, and alleviating the human and financial burdens associated with medication failure.

In traditional care settings, medications often fail to achieve their desired outcomes, with studies showing that a significant percentage of commonly prescribed drugs do not work as intended for many patients. For example, research indicates that antidepressants are effective in only about 50% of patients, leaving the other half, as in Emily's case, to endure prolonged symptoms while cycling through alternative treatments.[2] Similarly, blood pressure medications fail to adequately control hypertension in up to 40% of patients, increasing their risk of serious complications like stroke and heart attack.[3]

Cancer treatments also exhibit notable variability, with effectiveness rates often dependent on genetic and molecular factors. For some cancers, such as non-small-cell lung cancer, standard chemotherapy regimens yield meaningful responses in fewer than 30% of cases.[4] These statistics highlight the inefficiencies of traditional, one-size-fits-all approaches, particularly when dealing with conditions that demand precise interventions.

Trial-and-error prescribing is especially problematic for chronic conditions like diabetes, where achieving optimal blood sugar control may require months or years of experimenting with various medications and doses. Similarly, in asthma management, the variability in patient response to inhaled corticosteroids can lead to poorly controlled symptoms and frequent exacerbations. See Figure 5-1 for an illustration of the causes of medication failure.

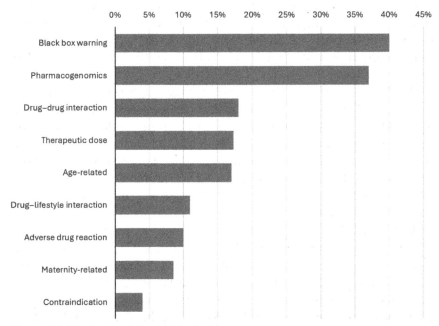

Figure 5-1: Medication failure risk by incidence rate
Source: Adapted from [5]

These inefficiencies contribute to patient frustration, delayed recovery, and higher costs for both individuals and the healthcare system. The failure to effectively manage these conditions can lead to increased hospitalizations, reduced productivity, and diminished quality of life. Addressing these challenges requires a shift to precision medicine, where treatments are selected based on a deeper understanding of a patient's unique biology, significantly improving the odds of therapeutic success and reducing the burdens of trial-and-error prescribing.

Adverse Drug Reactions

Another area of concern is adverse drug reactions (ADRs), a significant and preventable cause of morbidity and mortality in traditional care. According to the U.S. Food and Drug Administration (FDA), ADRs account for over two million hospitalizations alone in the United States each year.[6] Furthermore, ADRs are estimated to be the fourth leading cause of death in the United States, responsible for more than 100,000 fatalities each year—exceeding deaths from many chronic diseases.[7]

The human cost of ADRs is profound. Beyond hospitalizations, patients may suffer lasting harm, including chronic conditions resulting from organ damage or permanent disability. For example, medications prescribed for pain management, such as opioids, often result in dependency and overdose, leading to devastating individual and societal outcomes. Elderly populations are particularly vulnerable, with polypharmacy significantly increasing their risk of falls, cognitive decline, and other adverse events.

The financial toll of ADRs is equally staggering. Studies estimate that the direct cost of ADR-related hospitalizations exceeds $30 billion annually in the United States.[8] This figure does not account for the indirect costs, such as lost productivity, long-term disability, and caregiver burden, which further aggravate the economic impact. For employers, ADRs contribute to absenteeism and reduced workforce efficiency, whereas for the healthcare system, they add to the strain on emergency services and intensive care units.

Addressing ADRs requires a paradigm shift in how medications are prescribed and monitored. Precision medicine, which tailors treatment to individual genetic, environmental, and lifestyle factors, holds promise in significantly reducing ADRs. By identifying patients at higher risk for adverse reactions and selecting medications accordingly, precision medicine can improve patient safety, lower healthcare costs, and enhance overall treatment efficacy, marking a critical step forward in modern healthcare.

Adherence Challenges

Nonadherence to prescriptions is also a major factor in medication failure, leading to preventable disease progression, hospitalizations, and significant healthcare costs. Several factors contribute to nonadherence. Side effects, such as muscle pain from statins and nausea from cancer treatments, frequently discourage continued use. Complex regimens, involving multiple medications with varying schedules, increase the likelihood of missed doses or errors, particularly in conditions like diabetes or hypertension. Additionally, a lack of perceived efficacy often undermines adherence; patients who don't experience immediate improvement or fail to understand the long-term benefits may stop taking their medications.

The consequences are profound. Poor adherence accelerates disease progression in conditions like heart disease and diabetes, resulting in avoidable complications and hospitalizations. The financial impact is significant, with nonadherence estimated to cost the U.S. healthcare system over $100 billion annually.[9]

Addressing adherence challenges requires patient-centered solutions, including simplified regimens, better communication, and personalized interventions. Precision medicine can help optimize treatment plans and address individual barriers, reducing nonadherence and improving outcomes. By tailoring approaches to patients' needs, precision medicine offers a pathway to overcoming one of traditional care's greatest inefficiencies.

Cost of Medication Failure

The combined effects of medication failure impose a massive financial burden on healthcare systems and patients, with direct and indirect costs amounting to hundreds of billions of dollars annually. In the United States, nonoptimized medication use, such as prescribing ineffective drugs, poor adherence, and ADRs, accounts for over 16% of healthcare spending each year.[10] These readily avoidable costs stem from hospitalizations, emergency room visits, and repeated medical consultations required to manage complications or adjust treatment plans.

Public programs like Medicare and Medicaid bear a substantial share of this burden. Medicare, primarily comprised of the 60 million Americans over age 65, experiences substantial exposure through higher medication use that accompanies aging, and Medicaid programs face increased costs from complications in vulnerable populations. Private insurers are also impacted by inefficiencies, driving up premiums for employers and individuals alike. Figure 5-2 provides a perspective of the relative share of healthcare costs and impact of medication failure across the U.S. population.

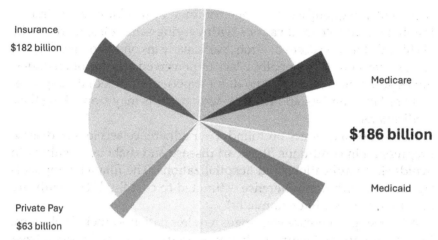

Insurance
$182 billion

Medicare

$186 billion

Medicaid

Private Pay
$63 billion

Figure 5-2: Annual cost of medication failure
Source: Adapted from [11]

The financial toll extends beyond direct healthcare expenses. Patients and families face hidden costs, including lost wages due to illness or caregiving responsibilities, as well as out-of-pocket expenses for additional treatments. The societal cost of diminished productivity is significant, particularly for working-age adults whose conditions are poorly managed due to medication failure. Moreover, medication failure erodes patients' quality of life. Chronic conditions left unmanaged due to ineffective or misused treatments can lead to pain, disability, and emotional distress. These intangible costs, although harder to quantify, amplify the overall burden on individuals and communities.

Precision medicine offers an opportunity to mitigate these financial and human costs. By aligning treatments with patients' unique biological profiles, precision medicine can reduce inefficiencies, prevent adverse events, and ensure that resources are directed toward effective care, delivering both economic and quality-of-life benefits. Unfortunately, medication efficacy is not uniform across all populations, with disparities arising from genetic, racial, and socioeconomic factors. Traditional prescribing methods, which often fail to consider these differences, aggravate inequities in healthcare outcomes and access.

Genetics play a significant role in how individuals metabolize medications.

Genetics play a significant role in how individuals metabolize medications. For example, the blood thinner warfarin has a narrow therapeutic window and is less effective or more dangerous for patients with specific genetic variations, particularly those of African or Asian descent. Similarly, asthma medications like long-acting beta-agonists have been shown to work less effectively in black patients, who also experience higher rates of asthma-related hospitalizations and deaths.

Socioeconomic factors compound these disparities. Patients in lower-income groups may have limited access to newer, more effective treatments or struggle with adherence due to cost, lack of education, or inadequate healthcare support. For instance, low-income individuals are more likely to use older, less-effective medications for chronic conditions like diabetes or hypertension, leading to poorer outcomes. The equity gap in traditional prescribing methods underscores the need for precision medicine, which considers genetic, environmental, and social determinants of health. By tailoring treatments to individual needs, precision medicine has the potential to close these gaps, improve efficacy across diverse populations, and advance health equity on a broader scale.

As we have seen, medication failure in traditional care is a costly and pervasive issue. With nearly half of patients experiencing ineffective treatments, ADRs cause millions of hospitalizations annually, and nonadherence adds billions of dollars in avoidable costs. These inefficiencies lead to preventable suffering, accelerated disease progression, and significant financial burdens for patients and healthcare systems alike.

The root causes—trial-and-error prescribing, inequities in drug efficacy, and system-wide inefficiencies—underscore the need for a paradigm-shifting approach. Precision medicine offers a path forward by tailoring treatments to individual genetic, environmental, and lifestyle factors, addressing the limitations of one-size-fits-all care. The next section explores how precision medicine eliminates trial-and-error prescribing, promising better outcomes, reduced costs, and greater equity in healthcare.

Precision Medicine's Approach: Eliminating Trial and Error

Trial-and-error prescribing remains a significant challenge in traditional healthcare. Patients often endure a lengthy and frustrating process of trying multiple medications before finding one that works, resulting in prolonged suffering, preventable complications, and wasted resources. For conditions like depression, hypertension, and diabetes, this iterative process can take months or even years, delaying effective treatment and increasing the risk of severe outcomes.

Precision medicine offers a pivotal solution by tailoring treatments to each patient's unique genetic, environmental, and lifestyle factors. Unlike the one-size-fits-all model, precision medicine uses advanced diagnostic tools and data to predict which medications will work best for a specific individual, minimizing the guesswork.

The overarching goal of precision medicine is to replace the outdated trial-and-error system with scientifically informed, patient-specific choices. By aligning treatments with the biological mechanisms of disease in each patient, precision medicine promises to improve efficacy, reduce adverse reactions, and optimize healthcare resources. This shift from reactive to proactive care not only enhances patient outcomes but also addresses inefficiencies that have long plagued the healthcare system, setting a new standard for how medications are prescribed and managed.

Personalized Care and Improved Outcomes

The personalized approach ensures that patients receive the right medication at the right dose from the outset, minimizing the uncertainty and frustration often associated with finding effective treatments.

One tangible advantage is improved treatment efficacy. For example, patients with HER2-positive breast cancer who receive targeted therapies like trastuzumab experience markedly better outcomes than those treated with conventional chemotherapy alone.

NOTE HER2-positive breast cancer is a type of breast cancer that has high levels of the protein *human epidermal growth factor receptor 2* (HER2).

Similarly, pharmacogenomic testing in patients with depression can identify antidepressants most likely to work for them, reducing the prolonged trial periods and improving response rates.

Precision medicine also enhances quality of life by reducing side effects. Traditional prescribing often exposes patients to medications that may not work and can cause debilitating adverse reactions. By selecting treatments suited to an individual's unique biology, precision medicine minimizes these risks, enabling patients to recover faster and with fewer complications.

Fewer side effects also contribute to better adherence to prescribed treatments. Patients who experience less discomfort and see tangible improvements are more likely to stay consistent with their medication regimens. This adherence leads to better long-term health outcomes, reducing complications, hospitalizations, and overall healthcare costs.

In this way, precision medicine also offers significant financial benefits for healthcare systems by addressing the inefficiencies inherent in traditional care. By reducing hospitalizations, minimizing ADRs, and eliminating the costly trial-and-error approach to prescribing, precision medicine has the potential to lower healthcare costs substantially while improving outcomes.

One key area of savings is the reduction in hospitalizations. Precision medicine can identify patients at higher risk for ADRs based on their genetic profiles, allowing clinicians to avoid prescribing medications likely to cause harm. Similarly, by tailoring treatments to the underlying biology of a disease, precision medicine reduces complications and the need for emergency interventions.

Fewer failed treatments naturally translates into cost savings as well. For example, pharmacogenomic testing for antidepressants has been shown to improve the time to find effective therapy by 53%, reducing unnecessary appointments, medication changes, and associated healthcare costs.[12] In oncology, targeted therapies for cancers like non-small-cell lung cancer have not only improved survival rates but also reduced the financial burden of ineffective chemotherapy regimens, demonstrating a measurable return on investment.

Public healthcare programs like Medicare and Medicaid stand to benefit significantly. Precision medicine can help optimize resource allocation by preventing hospital readmissions and reducing the need for high-cost treatments resulting from complications. For instance, in managing chronic conditions like diabetes and hypertension, precision approaches can prevent costly end-stage complications, saving billions annually. These savings can be reinvested into expanding access to advanced diagnostics and therapies, creating a virtuous cycle of cost-effectiveness and improved care.

In the long term, the systemic efficiencies generated by precision medicine contribute to a more sustainable healthcare model. By focusing on prevention, personalization, and optimized resource use, healthcare systems can better manage rising costs and aging populations. As precision medicine continues to evolve, its ability to align better outcomes with cost savings offers a compelling vision for the future of sustainable healthcare.

Barriers to Adoption

Despite its enormous potential, implementing precision medicine faces significant challenges. The cost of genetic testing and advanced diagnostic tools remains a major barrier. Although prices for sequencing technologies have decreased over time, the upfront expense can still be prohibitive for many patients and healthcare providers. Additionally, disparities in access to precision medicine disproportionately affect rural, low-income, and minority populations, aggravating existing inequities in healthcare.

Another challenge is the lack of clinician training in interpreting genetic data and integrating it into patient care. Many healthcare providers feel unprepared to use pharmacogenomic insights or recommend precision therapies, limiting adoption even when tools are available.

Initiatives aimed at overcoming these barriers are showing promise. Insurance coverage for pharmacogenomic testing is expanding, with some programs, including Medicare, reimbursing these services under certain conditions. Government-funded research programs like the National Institutes of Health's All of Us initiative are working to build diverse genetic datasets, paving the way for more equitable and accessible precision medicine solutions. Moreover, partnerships between private industry and public health organizations are driving innovation to lower costs and streamline the implementation of precision medicine.

Education and advocacy are critical to integrating precision medicine into mainstream healthcare. Training programs for clinicians and public awareness campaigns can help bridge knowledge gaps and promote broader acceptance of personalized approaches. By empowering both providers and patients with the tools and understanding needed to adopt precision medicine, these efforts can accelerate its integration and help ensure that its benefits are broadly accessible.

A Future Without Trial and Error

Such challenges notwithstanding, advancements in pharmacogenomics are poised to revolutionize medication management, driving precision medicine toward eliminating trial-and-error prescribing. Pharmacogenomics continues to refine our understanding of how genetic variations influence drug response, enabling clinicians to prescribe the most effective and safest medications from the outset. This science is rapidly expanding to cover a broader spectrum of medications and conditions, moving beyond behavioral health and pain management to include areas like cardiovascular disease, autoimmune disorders, and infectious diseases.

Envisioning a future healthcare system driven by precision medicine, trial-and-error prescribing could become largely obsolete. Patients would receive tailored treatments based on predictive models that consider their unique biology and lifestyle, reducing inefficiencies, improving outcomes, and transforming the patient experience. Precision medicine represents a fundamental shift, aligning healthcare delivery with the unique needs of every individual and setting a new standard for care.

By prioritizing evidence-based, individualized care, precision medicine holds the promise of a more effective and equitable healthcare system. As we move forward, it's valuable to reflect on the historical evolution of medicine. Past challenges and achievements provide context for the advancements shaping today's precision medicine revolution.

A HISTORICAL LOOK AT FALSE REMEDIES AND BREAKTHROUGHS IN MEDICATION

Imagine trusting a salesperson who promised their miracle elixir—made from the finest snake oil—would cure all ailments, from gout to baldness. Absurd? Yes. Effective? Absolutely not. Yet for centuries, remedies like this captivated people desperate for relief, reflecting humanity's enduring quest to conquer illness with a mix of ingenuity, desperation, and a pinch of gullibility.

Early medicine was as much trial and error as it was science. Treatments ranged from the mildly ineffective to the downright harmful, like using leeches to "balance the humors" or prescribing arsenic for stomach pains. Although laughable in hindsight, these efforts highlight the universal human desire to feel better, even when understanding of the human body was, shall we say, "under construction." Let's take a lighthearted yet enlightening journey through the evolution of medication.

The Era of Elixirs and Snake Oil

The 19th century was a golden age for patent medicines—although "golden" may better describe the profits than the products. These remedies, often sold as cure-alls, thrived in a time of maximal creativity and minimal regulation. From the streets of London to the wilds of America, peddlers of potions promised miracles in a bottle. Got rheumatism? Indigestion? A case of the vapors? There was an elixir for that—and it was probably laced with opium, alcohol, or a touch of arsenic.

Advertising was key to this boom, blending bold claims with captivating showmanship. Consider Radithor, a radioactive water marketed as "Perpetual Sunshine" for the body. Its creator touted it as a health tonic, but regular users discovered, painfully, that sipping radiation didn't cure ailments—it caused jaw decay and death (see Figure 5-3). Similarly, Cocaine Toothache Drops promised to soothe children's teething pains, which it did—until the addiction set in.

These "cures" often delivered unintended consequences far worse than the ailments they claimed to treat. Addiction, poisoning, and even death were frequent side effects, all while patent medicine companies thrived. Their secret? A lack of oversight. With no requirement to prove safety or efficacy, they mixed pseudoscience with marketing genius, ensuring that customers returned—if they survived.

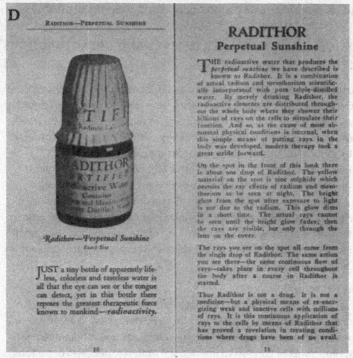

Figure 5-3: Radithor advertisement

Source: Blaufox et al., (2019) / with permission of Elsevier

Early Breakthroughs: When Guesswork Paid Off

Amid the sea of snake oil and radioactive water, though, a few accidental discoveries in medicine proved serendipitous—genuine breakthroughs that emerged from a mix of luck, curiosity, and sometimes sheer desperation. These moments, often born of guesswork, laid the foundation for modern medicine.

Take aspirin, for example. Derived from willow bark, its origins date back to ancient times when healers recommended chewing bark to relieve pain and fever. It wasn't until the late 19th century that chemists isolated salicylic acid and refined it into the pill we know today. A leap forward, to be sure—although the initial taste-testing phase must have left a few researchers grimacing.

Then there's penicillin, the poster child for accidental brilliance. In 1928, Alexander Fleming returned to his lab after vacation to find that mold had invaded his petri dishes. Instead of tossing them out, he noticed that the bacteria around the mold were dying. His curious observation gave the world its first antibiotic, saving countless lives and proving that sometimes a well-timed holiday can pay off.

Even nitroglycerin, infamous as an explosive, found a life-saving application. When physicians in the 19th century observed that workers exposed to nitroglycerin experienced relief from chest pain, they began experimenting with its medical use. The discovery of its vasodilating properties revolutionized the treatment of angina, although taking an explosive likely required a persuasive argument (see Figure 5-4).

I'm not gonna drink it. You drink it!

Figure 5-4: Discovery of medicinal nitroglycerin
Source: Generated with AI using DALL·E - OpenAI

These breakthroughs stood in stark contrast to the backdrop of dubious remedies peddled at the time. Quack cures offered quick fixes and empty promises, but these accidental discoveries underscored the power of observation, experimentation, and a willingness to question the ordinary.

The Quackery of Balms and Lotions

If you lived in the Victorian era and suffered from an ailment—be it a rash, joint pain, or even "nervous exhaustion"—there was likely a topical remedy for you. Balms, lotions, and ointments promised to cure everything, often with names as elaborate as their claims: "Dr. Euphoria's Magnetic Balm" or "The Universal Healing Salve." These concoctions, smeared onto unsuspecting skin, were marketed as cure-alls, but they often delivered more harm than healing.

Take mercury ointments, a favorite for treating skin conditions like syphilitic sores or acne. Although they did kill bacteria, they also poisoned the user, leading to hair loss, neurological issues, and, eventually, death. But hey, at least your rash cleared up.

Then there were the so-called "magnetic balms," which claimed to harness mysterious forces to draw out pain and illness. Advertisements boasted of their ability to "align the body's energies," but the only energy they likely aligned was that of the salesmen counting their profits. Often laced with camphor or menthol, these balms might have provided temporary cooling relief, but curing anything? Not so much (see Figure 5-5).

Figure 5-5: Magnetic balm

Source: From The Jeffersonian Democrat, Chardon, Ohio, Fri, Jul 27, 1860 / Public Domain

The fascination with topical remedies reflected a widespread belief that healing could be achieved externally, with little understanding of how the skin functioned as a barrier and organ. This lack of knowledge often led to treatments that were as damaging as they were ineffective. Over time, however, the study of dermatology and pharmacology advanced, recognizing the skin as a complex system that plays a vital role in overall health. Today, topical treatments are backed by rigorous science, from steroid creams to transdermal patches delivering precise doses of medication.

Modern Medicine's Missteps

Although modern medicine has brought us remarkable advancements, it's not without its stumbles—sometimes spectacular ones. The 20th century saw unfortunate missteps that remind us that even with lab coats and science, we're not immune to echoes of historical error.

Consider thalidomide, introduced in the late 1950s as a miracle sedative and treatment for morning sickness. Marketed as perfectly safe, it became a global sensation—until thousands of children were born with severe limb deformities (see Figure 5-6). The tragedy was a harsh lesson in the importance of testing and transparency.

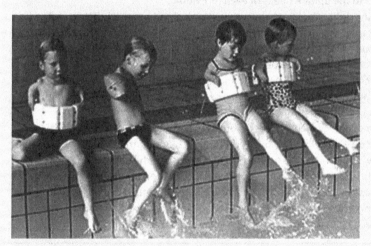

Figure 5-6: Children of thalidomide

Source: From "Thalidomide - Molecule of the Month". Licensed under Creative Commons License 4.0

Then there's fen-phen, the diet drug cocktail of the 1990s. Promising to melt pounds effortlessly, it became a blockbuster hit—until it was linked to heart valve damage and pulmonary hypertension. The fallout included lawsuits, shattered trust, and a lingering question: wasn't anyone suspicious that it was too good to be true?

Even early antidepressants, although groundbreaking, came with eyebrow-raising marketing. Advertisements in the 1960s and 1970s touted these drugs as cures for everything from sadness to "mother's little helper" fatigue, brushing off side effects like weight gain and emotional numbness as minor inconveniences. It was quackery in a suit and tie, delivered with a prescription pad.

These examples highlight how modern medicine, despite its advances, can veer into dangerous territory when shortcuts, exaggerated claims, or inadequate oversight take precedence. They also underscore the critical role of rigorous science, ethics, and regulation in safeguarding public health.

Lessons from History: A Foundation for Precision Medicine
History's tapestry of medical missteps and breakthroughs offers more than entertainment: it provides invaluable lessons that have shaped modern healthcare. From snake oil remedies to thalidomide tragedies, each failure has driven home the importance of rigor, ethics, and a deeper understanding of human biology. These lessons have laid the groundwork for today's advances in personalized medicine, where treatments are no longer "one size fits all" but tailored to the unique needs of each individual.

In hindsight, it's almost amusing to think about magnetic balms and radioactive tonics masquerading as cures. But these missteps remind us how far we've come. Today's precision medicine stands as a testament to the progress born from centuries of experimentation, both misguided and groundbreaking. By learning from history, we've transformed medicine into a science that combines compassion, evidence, and innovation, paving the way for a healthier, more personalized future.

History shows that progress is built on learning from mistakes, and precision medicine exemplifies this evolution. By tailoring treatments to individual biology, it embodies the lessons of centuries past while setting a course for the future. Still, it's fun to imagine future generations chuckling at today's practices—perhaps questioning why anyone thought swallowing cameras or tweaking gut microbes was cutting-edge. And so, the arc of medical progress continues, fueled by lessons, laughter, and innovation.

The Role of Pharmacogenomics in Precision Medicine

Pharmacogenomics lies at the heart of precision medicine, blending genetics and pharmacology to create a revolutionary approach to treatment. By decoding an individual's genetic makeup, pharmacogenomics ensures that the right drug is prescribed at the right dose from the start.

This section explores the significant role of pharmacogenomics in modern healthcare. First, we'll examine how genetic variations influence drug metabolism and efficacy, using real-world examples to illustrate its applications. Next, we'll delve into the technologies driving this field, such as genomic sequencing and bioinformatics.

Finally, we'll address the challenges and ethical considerations of integrating pharmacogenomics into everyday care, underscoring its potential to revolutionize medicine. By unlocking the genetic code of medication, pharmacogenomics is reshaping how we understand and practice healthcare.

The Science Behind Pharmacogenomics

Pharmacogenomics is revolutionizing medicine by revealing how genetic variations influence drug metabolism, efficacy, and toxicity. Central to this field is the understanding that individual genetic differences can determine whether a medication works effectively, causes side effects, or poses serious risks.

How an individual metabolizes a medication is determined jointly by *pharmacodynamics*, what the drug does to the body, and *pharmacokinetics*, what the body does to the drug. In short, a drug produces a certain effect on a person and that person's unique genetic makeup responds differently, sometimes significantly, than other people. Figure 5-7 illustrates this interaction between a drug and an individual.

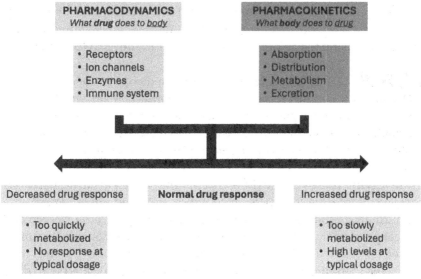

Figure 5-7: Pharmacodynamics and pharmacokinetics
Source: Adapted from [14]

A key player in this dance of drug metabolism is the family of cytochrome P450 enzymes, responsible for metabolizing approximately 70–80% of prescribed medications.[15]

NOTE Cytochrome P450 enzymes: A family of proteins that play a critical role in the metabolism of drugs, toxins, and endogenous compounds like hormones and lipids. Found primarily in the liver but also in other tissues, these enzymes are responsible for reactions that modify molecules, making them more water-soluble for easier excretion.

Variations in genes encoding these enzymes, such as CYP2D6 and CYP2C9, can significantly impact drug processing. For instance, individuals with certain CYP2D6 variants metabolize pain medications too quickly or too slowly, leading to either insufficient relief or dangerous toxicity. Similarly, genetic polymorphisms in CYP2C9 influence warfarin sensitivity, requiring careful dose adjustments to avoid bleeding complications.

NOTE CYP2D6 and CYP2C9 are cytochrome P450 enzymes that metabolize many drugs. CYP2D6 affects 25% of all drugs, including antidepressants and opioids, with genetic variations impacting efficacy; and CYP2C9 influences 15% of all drugs, including warfarin, NSAIDs, and antidiabetics, requiring careful management.[16]

Drug–gene interactions extend beyond metabolism to drug efficacy. For example, variations in the SLCO1B1 gene affect how statins are processed, increasing the risk of muscle-related side effects in some patients.

NOTE SLCO1B1 (Solute Carrier Organic Anion Transporter Family Member 1B1): Gene that encodes the OATP1B1 polypeptide protein, a liver-specific transporter responsible for moving drugs and endogenous compounds from the bloodstream for metabolism and excretion. This transporter plays a crucial role in the pharmacokinetics of various medications.[17]

Identifying these genetic differences allows clinicians to adjust doses or select alternative therapies, improving safety and effectiveness.

Pharmacogenomics bridges the gap between genomics research and clinical practice by translating genetic insights into actionable treatment strategies. Advances in genomic sequencing and bioinformatics now enable the identification of genetic polymorphisms associated with drug response. This knowledge is applied through pharmacogenomic testing, which guides medication choices based on a patient's genetic profile.

The implications are profound. By tailoring treatments to genetic predispositions, pharmacogenomics minimizes trial-and-error prescribing, reduces ADRs, and enhances therapeutic outcomes.

As a cornerstone of precision medicine, it turns the promise of genomics into practical solutions that improve patient care and advance the integration of personalized medicine into everyday healthcare.

One prominent application is in mental health care. Pharmacogenomic testing for genes like CYP2D6 and CYP2C19, which influence how medications like SSRIs and tricyclic antidepressants are metabolized, allows clinicians to identify the most effective drugs and optimal doses up front, significantly improving patient outcomes.

NOTE   CYP2C19 is another of the cytochrome P450 enzymes that metabolize many drugs. CYP2C19 affects approximately 10% of all drugs, including proton pump inhibitors, antidepressants, anti-epileptic drugs, and blood thinners.[18]

SSRIs (selective serotonin reuptake inhibitors) and tricyclic antidepressants are commonly prescribed medications used in the management of behavioral health.[19]

Pain management is another area where pharmacogenomics has made strides. Genetic testing for CYP2D6 helps identify individuals who metabolize opioids like codeine too rapidly or too slowly, reducing risks of toxicity or inadequate pain relief. This targeted approach ensures safer and more effective pain control for patients with diverse metabolic profiles.

In oncology, pharmacogenomics has revolutionized cancer treatment. Testing for the HER2 gene guides the use of targeted therapies like trastuzumab for HER2-positive breast cancer, significantly improving survival rates.[20] Similarly, genetic testing for TPMT activity helps determine the appropriate dosing of thiopurines in leukemia treatment, reducing toxicity while maintaining efficacy.

NOTE   TPMT: Thiopurine S-methyltransferase is an enzyme that metabolizes thiopurine drugs. Genetic variations affect drug efficacy and toxicity risk, particularly in chemotherapy and autoimmune treatments.[21]

Case studies also highlight the value of pharmacogenomics in preventing ADRs. For instance, patients with certain variants of the CYP2C19 gene are poor metabolizers of clopidogrel, an antiplatelet drug. Pharmacogenomic testing enables clinicians to prescribe alternative medications such as prasugrel, reducing the risk of blood clots and improving cardiovascular outcomes.

Pharmacogenomics in Adverse Drug Reactions

Adverse drug reactions are a significant issue in traditional care, affecting millions annually and imposing a substantial financial burden on healthcare systems. Pharmacogenomics offers a powerful solution by identifying individuals at risk of ADRs before treatment begins. By analyzing genetic variations that influence drug metabolism and immune responses, pharmacogenomics enables clinicians to tailor medications and dosages to minimize the likelihood of harmful side effects.

One notable example is the anticonvulsant carbamazepine, which can cause Stevens–Johnson syndrome (SJS) and toxic epidermal necrolysis (TEN): severe and potentially life-threatening skin reactions. Research has identified a strong association between the HLA-B*15:02 allele and an increased risk of SJS/TEN in individuals of Asian descent.

NOTE   HLA-B*15:02: Allele of the HLA-B gene, part of the human leukocyte antigen (HLA) system, which plays a critical role in immune response. This allele is strongly associated with severe, life-threatening hypersensitivity reactions to certain drugs and anticonvulsants.[22]

Pharmacogenomic testing for this allele allows clinicians to avoid prescribing carbamazepine to at-risk patients, dramatically reducing the incidence of these serious ADRs.

Another example involves warfarin, a common anticoagulant with a narrow therapeutic window. Variations in the CYP2C9 and VKORC1 genes affect how patients metabolize and respond to warfarin, increasing the risk of bleeding complications.

NOTE   Vitamin K Epoxide Reductase Complex Subunit 1 (VKORC1): Gene that encodes an enzyme essential for the vitamin K cycle, crucial for blood clotting. It is the primary target of commonly used anticoagulants.[23]

Pharmacogenomic testing helps adjust initial doses, improving safety and reducing the need for hospitalization due to overdosing. By reducing the incidence of ADRs, pharmacogenomics improves patient safety and alleviates financial pressures on the healthcare system. Fewer hospitalizations, emergency interventions, and long-term complications translate into significant cost savings. As pharmacogenomics becomes more widely adopted, its role in preventing ADRs underscores the transformative potential of precision medicine to deliver safer, more effective care.

Challenges to Implementation

Although pharmacogenomics holds immense promise, its integration into standard care faces significant challenges. For instance, the price of genetic testing has decreased over time, but it remains a barrier for many patients and healthcare systems, particularly in underfunded or resource-limited settings. Furthermore, limited insurance coverage for pharmacogenomic testing aggravates disparities in access.

Another challenge is the lack of clinician training. Many healthcare providers feel unprepared to interpret pharmacogenomic results or apply them in clinical decision-making. This knowledge gap slows adoption and reduces the potential benefits of precision medicine. Additionally, the lack of standardized protocols for testing and reporting creates inconsistencies in implementation.

Efforts to overcome these barriers are gaining momentum. Initiatives to incorporate pharmacogenomic data into electronic health records are streamlining its use in clinical workflows, enabling providers to access actionable insights during the prescribing process. Organizations like the Clinical Pharmacogenetics Implementation Consortium (CPIC) are developing guidelines to standardize testing and provide evidence-based recommendations for drug–gene interactions.[24]

Policy initiatives and expanded insurance coverage are also critical to broader adoption. Programs like Medicare's reimbursement for certain pharmacogenomic tests set an important precedent, encouraging other insurers to follow suit. These steps, combined with educational campaigns for clinicians and patients, are paving the way for pharmacogenomics to become an integral part of routine care, bridging the gap between scientific potential and everyday practice.

Pharmacogenomics in Population Health: A Broader Impact

Pharmacogenomics also has the potential to improve population health by addressing disparities in drug efficacy and safety across different groups. By tailoring medications to genetic profiles, pharmacogenomics ensures that treatments are optimized for individuals regardless of their ancestry, reducing inequities in healthcare outcomes. Pharmacogenomic testing can identify these variations, enabling clinicians to prescribe alternative therapies and improve outcomes for affected populations.

Ensuring equity in pharmacogenomics requires diverse genetic data in research. Historically, genomic studies have focused disproportionately on individuals of European ancestry, limiting the applicability of findings to other groups. Initiatives like the National Institutes of Health's All of Us program aim to address this gap by collecting genetic data broadly, ensuring that pharmacogenomic insights benefit all patients.[25]

> By incorporating diverse genetic data into drug development and clinical guidelines, pharmacogenomics can reduce health disparities on a societal level.

By incorporating diverse genetic data into drug development and clinical guidelines, pharmacogenomics can reduce health disparities on a societal level. The future of pharmacogenomics is poised to leverage other precision medicine technologies, too, reshaping how medications are prescribed and managed. AI-driven pharmacogenomic insights are one exciting frontier, with machine learning algorithms analyzing vast datasets to uncover complex drug–gene interactions and predict patient-specific responses. These insights can be integrated into electronic health records and prescribing tools, allowing clinicians to make highly informed decisions in real time.

Wearable devices also play a role in advancing pharmacogenomics. By collecting real-time health data, such as heart rate, blood pressure, and glucose levels, wearables provide dynamic feedback on how a patient responds to medications, further personalizing treatment plans. This integration can enable continuous monitoring and timely adjustments, enhancing both safety and efficacy.

Emerging therapies like gene editing may complement pharmacogenomics as well, by addressing the root causes of certain genetic variations. Technologies like CRISPR could potentially "correct" problematic genes, eliminating susceptibility to ADRs or ineffective treatments altogether.[26]

NOTE Clustered Regularly Interspaced Short Palindromic Repeats (CRISPR): Powerful gene-editing technology that allows scientists to precisely modify DNA in living organisms. It is based on a natural immune system found in bacteria, which use CRISPR-associated (Cas) enzymes to cut viral DNA.

Although still in its early stages, this convergence of pharmacogenomics and gene therapy offers profound possibilities.

Envisioning a future healthcare system, pharmacogenomics will certainly become a routine part of prescribing practices. Genetic testing at the point of care can help guide every prescription, reducing trial and error and ensuring safer, more effective treatments. This shift will not only improve individual outcomes but also enhance population health and optimize healthcare costs, marking a new era of precision medicine where pharmacogenomics is the standard, not the exception.

As we explore in the next section, pharmacogenomics stands out as a cornerstone of precision medicine, demonstrating how personalized approaches can drive better results and greater efficiency in healthcare delivery.

Comparison of Cost and Outcomes: Traditional vs. Precision Medicine

One of the most pressing questions in precision medicine is: How do the costs and outcomes of pharmacogenomics compare to traditional medicine? As healthcare systems grapple with rising expenses and uneven results, the stakes couldn't be higher. Pharmacogenomics offers the promise of improved efficacy, fewer ADRs, and optimized treatment plans. However, these benefits come with upfront costs, including genetic testing, specialized infrastructure, and clinician training.

On one hand, pharmacogenomics involves significant upfront investment—genetic testing, data infrastructure, and clinician training aren't cheap. But on the other hand, the potential long-term savings and improved outcomes are hard to ignore. Reducing ADRs, streamlining treatment selection, and minimizing trial-and-error prescribing could save billions of dollars, not to mention the priceless benefit of improving patients' lives.

This section dives into the numbers, examining both sides of the equation. We'll compare the initial costs of precision medicine with the downstream savings it generates, explore its impact on patient outcomes, and take a closer look at real-world examples where pharmacogenomics has delivered measurable benefits. Along the way, we'll address whether the promise of precision medicine can truly live up to its hype and, perhaps, learn a thing or two about what it means to balance innovation with practicality.

Economics of Traditional Medicine: Wasted Resources

Traditional medicine's reliance on trial-and-error prescribing creates significant inefficiencies, wasting both financial and human resources.

With just as many patients experiencing ineffective instead of effective treatments, the cost of these inefficiencies is staggering. Add in the expense of repeated doctor visits, and the financial burden grows exponentially. These inefficiencies not only inflate healthcare costs but also delay recovery, prolong suffering, and erode patient trust in the system, compounding the issue. Lost productivity due to untreated or poorly managed conditions burdens both patients and employers. Patients juggling multiple ineffective treatments often miss work or face decreased performance, impacting their financial stability and quality of life. Moreover, repeated treatment failures undermine patients' confidence in healthcare providers, reducing adherence to future regimens and worsening long-term outcomes.

These inefficiencies highlight the critical need for a shift toward more targeted, effective approaches. Precision medicine, with its ability to tailor treatments based on individual genetic and biological factors, addresses many of these pain points, promising to reduce waste, enhance outcomes, and rebuild trust in the healthcare system.

Cost and Outcomes of Pharmacogenomics

Implementing pharmacogenomics in precision medicine comes with significant upfront expenses. Although prices have decreased dramatically over the past decade, these tests still represent a notable financial hurdle, particularly for large-scale implementation.[27] Building the necessary infrastructure—such as data storage systems, bioinformatics tools, and integration with electronic health records—adds further costs. Additionally, clinician training to interpret and apply pharmacogenomic data is essential, requiring time and resources to ensure competency across the healthcare workforce.

The upfront costs of pharmacogenomics should be seen as an investment rather than an expense.

Despite these challenges, the upfront costs of pharmacogenomics should be seen as an investment rather than an expense. By reducing trial-and-error prescribing, preventing ADRs, and improving medication efficacy, pharmacogenomics promises significant long-term savings for both patients and healthcare systems. For example, avoiding a single hospitalization due to an ADR, averaging over $14,000 per admission, can offset the cost of genetic testing for multiple patients.[28] As these savings accumulate, pharmacogenomics demonstrates its value not only in improving outcomes but also in building a more efficient and equitable healthcare model.

Data illustrates the economic impact of this approach: ADRs, for example, can be substantially reduced through pharmacogenomic testing. Testing for the HLA-B*15:02 allele before prescribing carbamazepine, for instance, has nearly eliminated cases of Stevens–Johnson syndrome in certain populations, avoiding costly hospitalizations and improving patient outcomes.[29]

Cost-effectiveness comparisons underscore pharmacogenomics' value. As real-world evidence shows, the long-term savings and improved outcomes from pharmacogenomics far outweigh its initial expenses, identifying a pivotal catalyst for more efficient, patient-centered care.

Comparative Effectiveness

Actual applications of pharmacogenomics demonstrate its ability to improve outcomes and reduce costs compared to traditional approaches. The following examples highlight how precision medicine achieves faster recovery, fewer complications, and lower overall expenses.

Example 1: Warfarin Dosing

Warfarin, the widely used anticoagulant, has a limited therapeutic range, making accurate dosing critical. Traditional trial-and-error approaches often result in complications like bleeding or clotting, leading to hospitalizations. Pharmacogenomic studies identifying genetic contraindications revealed that patients who underwent testing had a 56% reduction in bleeding complications, with testing costs of several hundred dollars per patient offset by savings of $7,000 per patient through avoided complications.[30] These findings underscore how precision dosing improves safety while reducing healthcare costs.

Example 2: Clopidogrel and CYP2C19 Testing

Clopidogrel, an antiplatelet drug, is less effective in patients with certain CYP2C19 variants, increasing the risk of cardiovascular events; however, pharmacogenomic testing identifying poor metabolizers can enable physicians to prescribe alternatives like prasugrel. In one study, testing improved clinical outcomes with an average cost savings of $27,160 per patient due to reduced hospitalizations.[31] This precision approach not only saves lives but also alleviates the financial burden on healthcare systems.

Example 3: Cancer Therapies Tailored to Tumor Genetics

Targeted cancer therapies, such as trastuzumab for HER2-positive breast cancer, rely on genetic testing to match treatments to tumor profiles. Traditional chemotherapy often exposes patients to ineffective treatments with severe side effects. By identifying patients likely to respond, precision therapies improve survival rates and reduce wasted treatments. A comparative analysis showed that tailored therapy reduced the probability of ineffective treatment costs by as much as 26% and improved recovery times.[32]

These examples highlight how pharmacogenomics transforms care delivery. By reducing complications and improving efficacy, precision medicine achieves better patient outcomes and generates significant cost savings, redefining value in healthcare.

Societal Implications: The Broader Cost-Benefit

The widespread adoption of pharmacogenomics has the potential to generate systemic cost savings across the healthcare industry while improving societal health outcomes.

Integrating pharmacogenomics into public health initiatives is crucial to maximizing its cost-saving potential. Programs like the NIH's All of Us initiative ensure that pharmacogenomics benefits underserved and minority populations. Expanding insurance coverage and subsidizing testing in public health programs can further reduce disparities, making precision medicine accessible to all.

A common misconception about pharmacogenomics is that it is prohibitively expensive, deterring its adoption in routine care. Evidence from studies and pilot programs further dispels the cost barrier myth. For example, the study on warfarin dosing examined previously found that pharmacogenomic testing generated net savings within the first year of implementation. Similarly, the programs examined in mental health care demonstrated that genetic testing for antidepressant prescribing shortened treatment adjustment periods, reducing healthcare visits and improving productivity, with cost neutrality achieved within two to three years.

In the long term, pharmacogenomics promises not only to lower costs but also to improve productivity and societal health by enabling safer, more effective treatments for all patients. As a cornerstone of precision medicine, it represents a forward-looking investment in the sustainability and equity of healthcare systems.

What You Can Do

Pharmacogenomics offers a compelling value proposition, combining cost-effectiveness with improved patient outcomes. By reducing ADRs, hospitalizations, and ineffective treatments, it addresses the inefficiencies of traditional medicine and improves medication management, quickly offsetting costs with long-term savings and better health outcomes.

As a cornerstone of precision medicine, pharmacogenomics exemplifies the shift toward a more efficient, patient-centered healthcare system. It reduces waste, enhances treatment efficacy, and supports equitable care for diverse populations. So what does that mean for you—what can you do to put pharmacogenomics to work for your health today?

- *Create a personalized medication record.* Keep a record of your medications, dosage, and when they're taken. This information, combined with genetic insights, can help your doctor adjust prescriptions more precisely.

- *Use digital tools to track medication response.* Apps and wearable devices can help monitor how you respond to medications, tracking symptoms, side effects, and adherence. These insights, when shared with your doctor, can refine treatment strategies.

- *Familiarize yourself with drug actions and interactions.* All medications have known side effects, adverse event possibilities, and interactions with other drugs, which may affect how you metabolize them. Educating yourself about these interactions can help you make informed decisions about your treatment.

- *Ask your doctor about pharmacogenomic testing.* Many healthcare providers are now integrating genetic testing into medication management. Ask your physician whether pharmacogenomic testing could help personalize your prescriptions, especially if you've experienced ADRs or ineffective medications in the past.

- *Check with your insurance provider.* Coverage for pharmacogenomic testing is expanding. Contact your insurer to see if they cover genetic testing for medication management, particularly for conditions requiring long-term drug therapy.

- *Explore direct-to-consumer genetic testing.* If your doctor isn't familiar with pharmacogenomics, you can explore reputable direct-to-consumer genetic testing services that provide medication response reports. Be sure to choose one that offers clinically validated results and consult a healthcare professional to interpret them properly.

- *Advocate for precision medicine in your care.* If you're managing a chronic condition like hypertension, depression, or diabetes, discuss the possibility of precision medicine approaches with your healthcare team. Personalized medication selection can significantly reduce trial-and-error prescribing and improve outcomes.

■ *Encourage family members to consider testing.* Because genetics play a role in medication response, discussing pharmacogenomic testing with family members—especially those with similar medical conditions—can help them avoid ineffective or harmful medications.

■ *Stay up to date on advances in precision medicine.* Research in pharmacogenomics is rapidly evolving. Follow trusted sources like the National Institutes of Health or the Personalized Medicine Coalition to stay informed about new developments and emerging treatment options.

These steps empower you to integrate pharmacogenomics into your healthcare journey today, ensuring that your medications are working for you, not against you. By acting now, you can leverage the power of pharmacogenomics to optimize your care, minimize risks, and improve overall health outcomes, putting you in control of your treatment.

As research advances and access expands, users of pharmacogenomics will lead the charge in reshaping healthcare, benefiting ourselves as well as our families and future generations. By embracing this shift, you will not only improve your own health, but you will also be driving the movement toward smarter, safer, and more effective medical care.

Notes

1. Watanabe, Jonathan, McInnis, Terry, and Hirsch, Jan, "Cost of Prescription Drug–Related Morbidity and Mortality," *Annals of Pharmacotherapy* 52, no. 9 (2018): 829–837.

2. National Center for Biotechnology Information, "Depression: Learn More – How effective are antidepressants?," April 15 (2024), https://www.ncbi.nlm.nih.gov/books/NBK361016.

3. Calhoun, David, Booth, John, Oparil, Suzanne, Irvin, Marguerite, Shimbo, Daichi, Lackland, Daniel, Howard, George, Safford, Monika, and Muntner, Paul, "Refractory Hypertension: Determination of Prevalence, Risk Factors, and Comorbidities in a Large, Population-Based Cohort," *Hypertension* 63, no. 3 (2014): 451–458.

4. Sirohi, Bhawna, Ashley, Sue, Norton, Alison, Papadopoulos, Panagiotous, Priest, Kathryn, and O'Brien, Mary, "Early Response to Platinum-Based First-Line Chemotherapy in Non-small Cell Lung Cancer May Predict Survival," *Journal of Thoracic Oncology* 2, no. 8 (2007): 735–740.

5. Meyer, Tricia, "The Relevance of Black Box Warnings," Spring (2006), https://www.apsf.org/article/the-relevance-of-black-box-warnings.

Chanfreau-Coffinier, Catherine, Hull, Leland, Lynch, Julie, DuVall, Scott, Damrauer, Scott, Cunningham, Francesca, Voight, Benjamin, Matheny, Michael, Oslin, David, Icardi, Michael, and Tuteja, Sony, "Projected Prevalence of Actionable Pharmacogenetic Variants and Level A Drugs Prescribed Among US Veterans Health Administration Pharmacy Users," *JAMA Network Open* 2, no. 6 (2019): e195345.

Abdelkawy, Khaled, Kharouba, Maged, Shendy, Khloud, Abdelmagged, Omar, Galal, Naira, Tarek, Mai, Abdelgaied, Mohamed, Zakaria, Amr, and Mahmoud, Sherif, "Prevalence of Drug-Drug Interactions in Primary Care Prescriptions in Egypt: A Cross-Sectional Retrospective Study," *Pharmacy (Basel)* 11, no. 3 (2023): 106.

Garin, Noe, Sole, Nuria, Lucas, Beatriz, Matas, Laia, Moras, Desiree, Rodrigo-Troyano, Ana, Gras-Martin, Laura, and Fonts, Nuria, "Drug Related Problems in Clinical Practice: A Cross-Sectional Study on Their Prevalence, Risk Factors and Associated Pharmaceutical Interventions," *Scientific Reports* 11, no. 883 (2021): 883.

Caplan, Zoe, "U.S. Older Population Grew From 2010 to 2020 at Fastest Rate Since 1880 to 1890," May 25 (2023), https://www.census.gov/library/stories/2023/05/2020-census-united-states-older-population-grew.html.

Jones, Jeffrey, "Cigarette Smoking Rate Ties Last Year's Low," July 28 (2023), https://news.gallup.com/poll/648521/cigarette-smoking-rate-ties-year-low.aspx.

Ribeiro, Marisa, Motta, Antonio, Marcondes-Fonseca, Luiz, Kalil-Filho, Jorge, and Giavina-Bianchi, Pedro, "Increase of 10% in the Rate of Adverse Drug Reactions for Each Drug Administered in Hospitalized Patients," *Clinics (Sao Paulo)* 73 (2018): e185.

National Center for Health Statistics, "Life Expectancy in the U.S. Increased in 2021 for the First Time in Two Years," April 12 (2023), https://www.cdc.gov/nchs/pressroom/nchs_press_releases/2023/20230412.htm.

Rasool, Muhammad, Rehman, Anees, Khan, Irfanullah, Latif, Muhammad, Ahmad, Imran, Shakeel, Sadia, Sadiq, Muhammad, Hayat, Khezar, Shah, Shahid, Ashraf, Waseem, Majeed, Abdul, Hussain, Iltaf, and Hussain, Rabia, "Assessment of risk factors associated with potential drug-drug interactions among patients suffering from chronic disorders," *PLoS One* 18, no.1 (2023): e0276277.

6. Lazarou, Jason, Pomeranz, Bruce, and Corey, Paul. "Incidence of adverse drug reactions in hospitalized patients: a meta-analysis of prospective studies," *Journal of the American Medical Association* 279, no. 15 (1998): 1200–1205.

7. Ibid.

8. Ibid.

Aspinall, Sherrie, Vu, Michelle, Moore, Von, Jiang, Rong, Au, Anthony, Bounthavong, Mark, and Glassman, Peter, "Estimated Costs of Severe Adverse Drug Reactions Resulting in Hospitalization in the Veterans Health Administration," *JAMA Network Open* 5, no. 2 (2022): e2147909.

9. Iuga, Aurel, and McGuire, Maura, "Adherence and Health Care Costs," *Risk Management and Healthcare Policy* 7 (2014): 35-44.

10. Watanabe, et al., "Cost of Prescription Drug-Related Morbidity."

11. Centers for Medicare & Medicaid Services, "National Health Expenditure Fact Sheet," December 18 (2024), https://www.cms.gov/data-research/statistics-trends-and-reports/national-health-expenditure-data/nhe-fact-sheet.

Jarvis, et al., "Real-World Impact."

Fragala, Maren, Keogh, Murray, Goldberg, Steven, Lorenz, Raymond, and Shaman, Jeffrey, "Clinical and Economic Outcomes of a Pharmacogenomics-Enriched Comprehensive Medication Management Program in a Self-Insured Employee Population," *Pharmacogenomics Journal* 24, no. 5 (2024): 30.

12. Tanner, Julie-Anne, Davies, Paige, Overall, Christopher, Grima, Daniel, Nam, Julian, and Dechairo, Bryan, "Cost-Effectiveness of Combinatorial Pharmacogenomic Testing for Depression from the

Canadian Public Payer Perspective," *Pharmacogenomics* 21, no. 8 (2020): 521–531.

13. Ahmed, Shabbir, Zhou, Zhan, Zhou, Jie, and Chen, Shu-Qing, "Pharmacogenomics of Drug Metabolizing Enzymes and Transporters: Relevance to Precision Medicine," *Genomics, Proteomics & Bioinformatics* 14, no. 5 (2016): 298–313; as corrected.

14. Zanger, Ulrich, and Schwab, Matthias, "Cytochrome P450 Enzymes in Drug Metabolism: Regulation of Gene Expression, Enzyme Activities, and Impact of Genetic Variation," *Pharmacology & Therapeutics* 138, no. 1 (2013): 103–141.

15. Belle, Donna, and Singh, Harleen, "Genetic Factors in Drug Metabolism," *American Family Physician* 77, no. 11 (2008): 1553–1560.

   Van Booven, Derek, Marsh, Sharon, McLeod, Howard, Carrillo, Michelle, Sangkuhl, Katrin, Klein, Teri, and Altman, Russ, "Cytochrome P450 2C9-CYP2C9," *Pharmacogenetics and Genomics* 20, no. 4 (2010): 277–281.

16. Häkkinen, Katja, Kiander, Wilma, Kidron, Heidi, Lähteenvuo, Markku, Urpa, Lea, Lintunen,. Jonne, Vellonen, Kati-Sisko, Auriola, Seppo, Holm, Minna, Lahdensuo, Kaisla, Kampman, Olli, Isometsä, Erkki, Kieseppä, Tuula, Lönnqvist, Jouko, Suvisaari, Jaana, Hietala, Jarmo, Tiihonen, Jari, Palotie, Aarno, Ahola-Olli, Ari, and Niemi, Mikko, "Functional Characterization of Six SLCO1B1 (OATP1B1) Variants Observed in Finnish Individuals with a Psychotic Disorder," *Molecular Pharmaceutics* 20, no. 3 (2023): 1500–1508.

17. MedlinePlus, "CYP2C19 Gene," December 1 (2015), https:// medlineplus.gov/genetics/gene/cyp2c19.

18. Anderson, Ian, "Selective Serotonin Reuptake Inhibitors Versus Tricyclic Antidepressants: A Meta-Analysis of Efficacy and Tolerability," *Journal of Affective Disorders* 58, no. 1 (2000): 19–36.

19. Johnson, Kai, Ni, Ai, Quiroga, Dionisia, Pariser, Ashley, Sudheendra, Preeti, Williams, Nicole, Sardesai, Sagar, Cherian, Mathew, Stover, Daniel, Gatti-Mays, Margaret, Ramaswamy, Bhuvaneswari, Lustberg, Maryam, Jhawar, Sachin, Skoracki, Roman, and Wesolowski, Robert, "The Survival Benefit of Adjuvant Trastuzumab With or Without Chemotherapy in the Management of Small (T1mic, T1a, T1b, T1c), Node Negative HER2+ Breast Cancer," *NPJ Breast Cancer* 10, no. 1 (2024): 49.

20. Coulthard, Sally, and Hall, Andy, "Recent Advances in the Pharmacogenomics of Thiopurine Methyltransferase," *Pharmacogenomics Journal* 1, (2001): 254–261.

21. Amstutz, Ursula, Shear, Neil, Rieder, Michael, Hwang, Soomi, Fung, Vincent, Nakamura, Hidefumi, Connolly, Mary, Ito, Shinya, and Carleton, Bruce, "Recommendations for HLA-B*15:02 and HLA-A*31:01 Genetic Testing to Reduce the Risk of Carbamazepine-Induced Hypersensitivity Reactions," *Epilepsia* 55, no. 4 (2014): 496–506.

22. Rost, Simone, Fregin, Andreas, Ivaskevicius, Vytautas, Conzelmann, Ernst, Hörtnagel, Konstanze, Pelz, Hans-Joachim, Lappegard, Knut, Seifried, Erhard, Scharrer, Inge, Tuddenham, Edward, Müller, Clemens, Strom, Tim, and Oldenburg, Johannes, "Mutations in VKORC1 Cause Warfarin Resistance and Multiple Coagulation Factor Deficiency Type 2," *Nature* 427, no. 6974 (2004): 537–541.

23. Clinical Pharmacogenetics Implementation Consortium, "CPIC Guidelines," February 3 (2025), https://cpicpgx.org/guidelines.

24. All of Us Research Program, "Program Overview."

25. Gostimskaya, Irina, "CRISPR-Cas9: A History of Its Discovery and Ethical Considerations of Its Use in Genome Editing," *Biochemistry (Moscow)* 87, no. 8 (2022): 777–788.

26. Colby, "Whole Genome Sequencing Cost."

27. Sultana, et al., "Clinical and Economic Burden."

28. Chen, Pei, Lin, Juei-Jueng, Lu, Chin-Song, Ong, Cheung-Ter, Hsieh, Peiyuan, Yang, Chih-Chao, Tai, Chih-Ta, Wu, Shey-Lin, Lu, Cheng-Hsien, Hsu, Yung-Chu, Yu, Hsiang-Yu, Ro, Long-Sun, Lu, Chung-Ta, Chu, Chun-Che, Tsai, Jing-Jane, Su, Yu-Hsiang, Lan, Sheng-Hsing, Sung, Sheng-Feng, Lin, Shu-Yi, Chuang, Hui-Ping, Huang, Li-Chen, Chen, Ying-Ju, Tsai, Pei-Joung, Liao, Hung-Ting, Lin, Yu-Hsuan, Chen, Chien-Hsiun, Chung, Wen-Hung, Hung, Shuen-Iu, Wu, Jer-Yuarn, Chang, Chi-Feng, Chen, Luke, Chen, Yuan-Tsong, and Shen, Chen-Yang, "Carbamazepine-Induced Toxic Effects and HLA-B*1502 Screening in Taiwan," *New England Journal of Medicine* 364, no. 12 (2011): 1126–1133.

29. Hillman, Michael, Wilke, Russell, Yale, Steven, Vidaillet, Humberto, Caldwell, Michael, Glurich, Ingrid, Berg, Richard, Schmelzer, John, and Burmester, James, "A Prospective, Randomized Pilot Trial of

Model-Based Warfarin Dose Initiation Using CYP2C9 Genotype and Clinical Data," *Clinical Medicine & Research* 3, no. 3 (2005): 137–145.

Jarvis, et al., "Real-World Impact."

30. Reese, Emily, Mullins, Daniel, Beitelshees, Amber, and Onukwugha, Eberechukwu, "Cost-Effectiveness of Cytochrome P450 2C19 Genotype Screening for Selection of Antiplatelet Therapy with Clopidogrel or Prasugrel," *Pharmacotherapy* 32, no. 4 (2012): 323–332.

31. Shi, Demin, Liang, Xueyan, Li, Yan, and Chen, Lingyuan, "Cost-Effectiveness of Trastuzumab Deruxtecan for Previously Treated HER2-Low Advanced Breast Cancer," *PLoS One* 18, no. 8 (2023): e0290507.

Through the Lens Clearly

Medicine has always been a journey of trial and discovery. From ancient herbal remedies and spiritual healing practices to the advent of surgery and the germ theory of disease, each era of medicine has been defined by its understanding of the human body and its afflictions. Early civilizations experimented with treatments often rooted in superstition or limited observation, laying the groundwork for centuries of incremental progress. The Renaissance brought anatomical studies and scientific methods, unveiling a clearer picture of how the body functions. By the 19th and 20th centuries, breakthroughs like antibiotics, vaccines, and imaging technologies transformed our ability to combat disease.

Despite these advancements, much of medicine has relied on generalized approaches, treating patients based on population averages rather than individual variability. The rise of genetics, molecular biology, and data-driven insights in recent decades has radically shifted this paradigm. Today, we understand that factors such as a person's DNA, environment, and lifestyle significantly influence their health outcomes.

This chapter offers a historical lens revealing how medicine's past challenges and achievements have shaped our present understanding. By examining this evolution, we can appreciate how precision medicine—

tailored to the unique characteristics of each individual—is not merely a trend but the next logical step in the age-old pursuit of healing. It represents the culmination of centuries of striving to understand and treat the human body with greater specificity and success.

Assembling the Pieces: The History of Medical Innovation

The origins of medicine are deeply rooted in the interplay between mysticism and empiricism. Early civilizations approached healing through shamanism, where spiritual leaders acted as intermediaries between the natural and supernatural worlds. Shamans used rituals, chants, and symbols to expel illnesses believed to be caused by malevolent spirits or divine displeasure. Simultaneously, practical remedies, such as herbal treatments derived from observation and trial, began to emerge. Indigenous communities around the globe developed knowledge of medicinal plants, forming the foundation of pharmacology.

In ancient India, Ayurvedic medicine blended natural observation with metaphysical beliefs, emphasizing balance among bodily elements: Vata, Pitta, and Kapha. Similarly, traditional Chinese medicine (TCM) relied on concepts like Qi (life force) and Yin–Yang balance while using acupuncture and herbal concoctions to restore harmony (see Figure 6-1). These systems represented an early attempt to categorize and address human health based on both empirical observation and spiritual frameworks.

Organized medical records began to take shape in ancient Egypt and Mesopotamia, marking a significant step toward structured healthcare. Egyptian papyri, such as the Edwin Smith Papyrus (circa 1600 BCE), documented diagnoses, treatments, and surgical techniques. In Mesopotamia, clay tablets recorded prescriptions, symptoms, and prognoses, reflecting an advanced understanding of diseases within the context of their religious and cultural beliefs. This blending of mysticism and empiricism laid the groundwork for future medical advancements. By recording their findings and refining their methods, ancient practitioners began the long journey toward disciplined, evidence-based medicine, illuminating how early practices informed the structure and philosophy of healthcare for millennia.

The transition from superstition to science-based reasoning in medicine, however, owes much to Hippocrates, often called the "Father of Medicine." Living in ancient Greece around the 5th century BCE,

Hippocrates challenged the prevailing belief that illness was a punishment from the gods. Instead, he emphasized that diseases arose from natural causes and could be understood through careful observation and logical analysis. His principles laid the foundation for a rational approach to medicine, focusing on diagnosing symptoms, monitoring progression, and prescribing treatments based on empirical evidence.

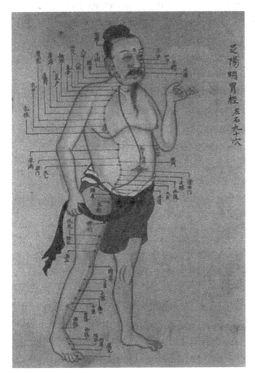

Figure 6-1: Early Chinese medical chart with acupuncture points
Source: Unknown / Wikimedia Commons / Public domain

Hippocratic principles extended beyond clinical methods to encompass ethics, as embodied in the enduring Hippocratic Oath. This framework established medicine as a moral practice, advocating for the primacy of patient welfare—a philosophy that still underpins medical professionalism today.

NOTE Hippocratic Oath

I swear by Apollo Healer, by Asclepius, by Hygieia, by Panacea, and by all
the gods and goddesses, making them my witnesses, that I will carry out,
according to my ability and judgment, this oath and this indenture.

To hold my teacher in this art equal to my own parents; to make him partner in my livelihood; when he is in need of money to share mine with him; to consider his family as my own brothers, and to teach them this art, if they want to learn it, without fee or indenture; to impart precept, oral instruction, and all other instruction to my own sons, the sons of my teacher, and to indentured pupils who have taken the Healer's oath, but to nobody else.

I will use those dietary regimens which will benefit my patients according to my greatest ability and judgment, and I will do no harm or injustice to them. Neither will I administer a poison to anybody when asked to do so, nor will I suggest such a course. Similarly I will not give to a woman a pessary to cause abortion. But I will keep pure and holy both my life and my art. I will not use the knife, not even, verily, on sufferers from stone, but I will give place to such as are craftsmen therein.

Into whatsoever houses I enter, I will enter to help the sick, and I will abstain from all intentional wrong-doing and harm, especially from abusing the bodies of man or woman, bond or free. And whatsoever I shall see or hear in the course of my profession, as well as outside my profession in my intercourse with men, if it be what should not be published abroad, I will never divulge, holding such things to be holy secrets.

Now if I carry out this oath, and break it not, may I gain for ever reputation among all men for my life and for my art; but if I break it and forswear myself, may the opposite befall me.[1]

Building on Hippocrates' work, the Roman physician Galen (2nd century CE) significantly influenced Western medicine. Galen's meticulous studies of anatomy, based largely on animal dissections, advanced understanding of bodily structures and functions. However, his reliance on incorrect assumptions—such as the notion that blood originated in the liver and was consumed as fuel—led to enduring misconceptions in medicine. Galen's writings dominated medical thought for over a millennium, often impeding progress, as they were treated as infallible.

Together, Hippocrates and Galen shaped the trajectory of Western medicine. Whereas Hippocrates introduced an organized, ethical approach, Galen's contributions underscore the importance—and limitations—of early medical inquiry in advancing human health.

Medieval Medicine: Faith, Plague, and Survival

The medieval period that followed in Europe was marked by a return to the intertwining of the supernatural and medicine. Diseases were often seen as divine punishment or tests of faith, and healing practices frequently involved prayer, relics, and rituals performed by clergy. Monasteries served as centers of medical knowledge, preserving ancient texts and offering rudimentary care; and like the charitable cloisters of St. Giles, St. Anthony, and St. Leonard, they served the spiritual, hospitality, and research needs of the community. The term *hospital* evolved naturally from the mission of these early establishments; however, the dominance of spiritual interpretation often limited the advancement of empirical medical practices in Europe during this era.

NOTE Hildegard of Bingen, a German Benedictine abbess and polymath active as a medical writer and practitioner during the High Middle Ages, was typical of the imprecise influence of religion on medicine:

Let a man who has an overabundance of lust in his loins cook wild lettuce in water and pour it over himself in a sauna.[2]

In contrast, the Islamic world flourished as a hub of medical innovation, blending knowledge from Greek, Roman, Persian, and Indian sources. One of the most influential figures was Avicenna (Ibn Sina), whose Canon of Medicine, completed in 1025, became a cornerstone of medical education for centuries.[3] Avicenna organized medical knowledge, emphasized empirical observation, and explored topics such as contagious diseases, pharmacology, and the importance of hygiene. His work advanced both theoretical understanding and practical applications, setting a high standard for medical scholarship.

The arrival of the Black Death in the 14th century changed things, however, devastating Europe and killing an estimated 50 million people; almost 50% of the population of England alone was decimated (see Figure 6-2).[4] It was this pandemic that exposed the limitations of medieval medicine, as prayers and religious rituals failed to halt the spread of the disease. The scale of the plague prompted efforts to understand its causes, leading to the development of rudimentary public health measures such as quarantines and improved sanitation. Although rooted in fear and desperation, these measures marked an early recognition of disease transmission and the importance of collective action in health crises.

The medieval era, defined by faith and survival, underscored both the stagnation of dogmatic approaches and the necessity of innovation in the face of calamity. It laid the groundwork for a shift toward more scientific approaches in the centuries that followed.

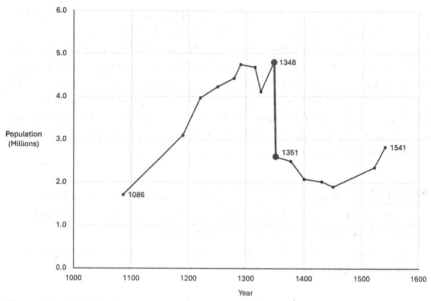

Figure 6-2: Black Death impact on the English population, 1348 to 1351
Source: Adapted from [5]

The Renaissance and Revolution: Foundation of Modern Medicine

The Renaissance marked a turning point in the history of medicine as the rediscovery of ancient Greek and Roman texts, coupled with a burgeoning curiosity about the natural world, revolutionized medical thought. This period saw a break from reliance on tradition, ushering in a renewed focus on empirical observation and scientific inquiry. Andreas Vesalius, a pioneering anatomist of the 16th century, was at the forefront of this shift. His seminal work, *De humani corporis fabrica* (On the Fabric of the Human Body), challenged longstanding misconceptions by providing detailed anatomical illustrations based on human dissection rather than animal models.[6] Vesalius's findings corrected errors perpetuated since Galen, advancing the understanding of human anatomy.

Building on these advancements, English physician William Harvey made a groundbreaking discovery in the 17th century: the circulatory system. Through meticulous experiments and observation, Harvey demonstrated that the heart functions as a pump, circulating blood through a closed system of arteries and veins. This insight not only overturned centuries of medical dogma but also exemplified the power of the scientific method in uncovering biological truths.

The Renaissance and subsequent Scientific Revolution also laid the groundwork for observational science and early clinical trials. Physicians began rigorously documenting patient cases, identifying patterns, and testing treatments in controlled ways. For example, James Lind's mid-18th century experiments with citrus fruits to combat scurvy among sailors highlighted the efficacy of controlled trials in establishing effective therapies.

This era's emphasis on observation, experimentation, and critical evaluation laid the foundations of modern medicine. By embracing scientific methods, it bridged the gap between ancient wisdom and contemporary medical practices, setting the stage for the significant discoveries of later centuries.

The Industrial Era that followed marked an era of extraordinary innovation, where technological advancements intersected with medical science to transform healthcare. Instruments such as the microscope, refined by Antonie van Leeuwenhoek and later improved during the 19th century, unveiled the previously unseen world of microorganisms, providing crucial insights into human biology and disease. The invention of the stethoscope by René Laennec in 1816 revolutionized diagnostic practices, allowing physicians to listen directly to the workings of the heart and lungs. By the late 19th century, Wilhelm Röntgen's discovery of X-rays offered a noninvasive method to visualize internal structures, revolutionizing diagnostics and surgical planning.

NOTE Auscultation: The act of listening to internal body sounds such as the heart, lungs, and intestines using a stethoscope or, in some cases, directly with the ear; a fundamental diagnostic technique in clinical medicine.

Concurrently, the germ theory of disease emerged, fundamentally reshaping the understanding of infection. Louis Pasteur's experiments with wine fermentation in the 1860s demonstrated that microorganisms were the agents of spoilage and disease, leading to innovations such as pasteurization and vaccination. Building on Pasteur's work, Robert Koch identified specific pathogens responsible for diseases like tuberculosis and cholera, establishing a scientific basis for combating infectious diseases. These breakthroughs spurred public health initiatives, including sanitation reforms and vaccination campaigns, dramatically reducing mortality rates.

The Industrial Era also saw the professionalization of medicine through organized education and standardization. In the United States, the Flexner Report of 1910 reformed medical education by emphasizing rigorous scientific training and clinical practice, setting a precedent for modern medical schools.[7] Licensing and regulatory bodies emerged, ensuring that practitioners met established standards of knowledge and competence.

This era's technological and theoretical advancements laid the groundwork for the modern medical landscape. By combining innovation with organized education and professionalization, the Industrial Era transformed medicine into a more effective and accessible discipline, capable of addressing the complexities of human health.

The 20th Century: Medicine at a Crossroads

The 20th century was a rapidly advancing period for medicine, marked by groundbreaking discoveries, the emergence of specialized disciplines, significant public health achievements, and a growing focus on medical ethics. These developments collectively reshaped the practice of medicine and its role in society.

One of the century's most pivotal breakthroughs was the discovery and widespread use of antibiotics, beginning with Alexander Fleming's identification of penicillin in 1928. Antibiotics revolutionized the treatment of bacterial infections, saving millions of lives and extending life expectancy worldwide. (See Figure 6-3 for an illustration of the timeline of antibiotics development.) Similarly, vaccines advanced dramatically, with the eradication of smallpox in 1980 standing as a monumental public health milestone. These developments ushered in the modern field of pharmacology, fostering the development of targeted drugs for an array of diseases.

Specialization became another defining feature of 20th century medicine. Disciplines such as cardiology, oncology, and neurology emerged, allowing for more focused and effective treatment of complex conditions. Modern surgery advanced with innovations in anesthesia, aseptic techniques, and tools such as laparoscopes, enabling previously unimaginable procedures. The advent of medical imaging technologies, including computed tomography (CT) scans and magnetic resonance imaging (MRI), further revolutionized diagnostics, providing unparalleled insights into the human body.

However, the century was also marked by ethical challenges that reshaped medical practice. The Tuskegee Syphilis Study, which withheld treatment from black men to observe the natural progression of the disease, was a glaring example of medical exploitation.[9] This, along with revelations of unethical experiments during World War II, led to the establishment of the Nuremberg Code, a set of principles emphasizing informed consent and the protection of human subjects in research. These milestones underscored the necessity of prioritizing patient rights and ethical considerations in medical care.

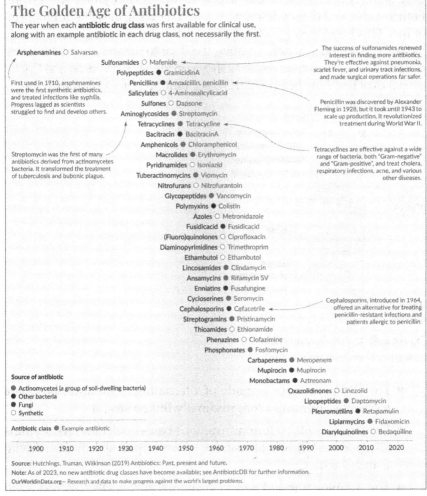

Figure 6-3: The Golden Age of antibiotics

Source: [8]/Our World in Data/CC BY 4.0

As medicine entered the 21st century, these advancements and lessons formed the foundation for a more sophisticated and humane healthcare system. The 20th century's legacy is one of both unprecedented progress and a renewed commitment to ethical practice, shaping the future of medicine as it navigates new challenges and opportunities.

The Digital Revolution and Genomics

The Digital Revolution and advancements in genomics have ushered medicine into an unprecedented era of precision and accessibility. With the discovery of DNA's double-helix structure in 1953, James Watson and Francis Crick laid the foundation for understanding the genetic blueprint of life. This breakthrough changed medicine by revealing how genetic variations influence health, opening the door to diagnosing and treating diseases at their molecular roots. As shown in Figure 6-4, Watson and Crick's model explained the beauty of the basic building block of human life:

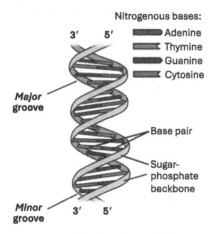

Figure 6-4: The DNA double helix
Source: [10] / Rice University / CC BY 4.0

- The DNA backbone is made of alternating sugar and phosphate molecules, forming a longitudinally linked strand.

- The strand contains four nitrogenous bases—adenine (A), thymine (T), guanine (G), and cytosine (C)—which pair specifically: A with T and G with C.

- The double-stranded helical structure makes DNA robust: each strand a complementary image of the other, enabling replication, stability, and error correction.

- DNA enables the storage, transmission, and expression of genetic information, ensuring faithful replication across generations.

Beyond its structure, DNA also explained the durable process of transmission of genetic information—and health—across generations:

- *Meiotic division*, creating reproductive cells bearing one-half of the parent's DNA.

- *Fertilization*, with sperm and egg merging to form a unique genetic blueprint and pairing of DNA strands.
- *Mitotic division*, rapid cell replication enabling growth and specialization.

These discoveries paved the way for genomic sequencing, marking a pivotal leap in medical science. Initiatives like the Human Genome Project (HGP), completed in 2003, decoded the entire human genome, identifying the genetic instructions that govern human biology. This monumental achievement catalyzed personalized medicine, enabling clinicians to tailor treatments based on a patient's genetic profile. Advances in sequencing technology have since made genomic testing faster, more affordable, and widely accessible, revolutionizing fields such as oncology, where treatments can now target specific mutations in tumors.

Concurrently, the rise of digital technology has transformed the infrastructure of healthcare delivery. Electronic health records replaced paper charts, providing centralized, accessible, and secure repositories for patient data. This shift has streamlined communication between healthcare providers and improved continuity of care. Meanwhile, telemedicine has bridged geographical barriers, allowing patients to consult with specialists remotely. The COVID-19 pandemic further accelerated the adoption of telehealth, making it an essential component of modern healthcare.

Wearable technology contributes as a key player in the digital health landscape. Devices like smartwatches and fitness trackers collect real-time data on metrics such as heart rate, sleep patterns, and physical activity. These insights empower individuals to monitor their health proactively while providing clinicians with valuable data for personalized care.

Together, the integration of genomics and digital technologies has redefined medicine, shifting the focus from reactive treatment to proactive prevention and precision. The present era of medicine is defined by groundbreaking innovations that are bringing this integration of precision medicine closer to reality. Technologies such as CRISPR, the revolutionary gene-editing tool, have enabled scientists to modify DNA with unprecedented precision.

This has opened doors to potential cures for genetic disorders such as sickle cell anemia and holds promise for combating complex diseases like cancer. Similarly, artificial intelligence (AI) is transforming diagnostics. Machine learning algorithms can analyze medical images, genomic data, and patient histories with remarkable accuracy, providing faster and more precise diagnoses than ever before.

The history of medical innovation offers hope and perspective. Each milestone—whether the discovery of antibiotics, the development of vaccines, or the decoding of the human genome—has overcome obstacles to achieve substantial impact. Precision medicine represents the next logical step in this continuum, building on centuries of progress to deliver more individualized and effective care.

As we navigate this new frontier, the lessons of history remind us that innovation thrives when guided by a commitment to equity, ethics, and the pursuit of better health. By addressing current challenges, precision medicine has the potential to not only revolutionize healthcare but to do so in a way that is inclusive and sustainable, ensuring its benefits reach everyone.

Implementing the Insights: From Bloodletting to Precision Medicine

Bloodletting, once a cornerstone of medical practice, epitomizes the outdated, one-size-fits-all approach to healthcare. For centuries, physicians relied on this technique to treat a variety of ailments, from fevers to mental illness, believing it could restore balance to the body's humors. Despite its ubiquity, bloodletting often caused more harm than good, reflecting the limitations of a medical system that lacked the tools to understand the complexity of individual health.

In stark contrast, modern medicine is defined by precision and customization. Advances in genomics, molecular biology, and data analytics have ushered in an era where treatments can be tailored to the unique genetic, environmental, and lifestyle factors of each patient. Technologies like CRISPR, AI-driven diagnostics, and wearable devices empower physicians to make targeted decisions, reducing risks and improving outcomes. This shift from generalized to personalized care represents a fundamental change in how we approach health and healing.

This section traces the key medical advancements that paved the way to precision medicine. From the crude early practices to the sophisticated tools of today, the journey highlights how each breakthrough has contributed to a deeper understanding of the human body and the diseases that affect it. By exploring this evolution, we can appreciate how the lessons of the past have shaped the innovative, patient-centered future of healthcare.

"Heroic" Medicine

For centuries, bloodletting was a cornerstone of medical practice, rooted in Galen's theory of humors. Developed in ancient Greece and widely adopted by medieval and Renaissance physicians, the humoral theory posited that health depended on the balance of four bodily fluids: blood, phlegm, yellow bile, and black bile. Illness, it was believed, resulted from an imbalance, and bloodletting—often through cuts or the application of leeches—was thought to restore equilibrium. Despite its widespread use, the practice was rarely effective and often harmful, weakening patients rather than curing them.

Bloodletting was just one of several early medical practices driven more by tradition than evidence. *Trepanation*, the practice of drilling holes into the skull, was used to treat conditions ranging from epilepsy to mental illness, likely based on the belief that it released evil spirits. Purging, which involved inducing vomiting or diarrhea, aimed to expel perceived toxins but often left patients dehydrated and malnourished. These practices, characterized by aggressive, often harmful interventions, came to be known as "heroic" medicine and persisted for centuries, sustained by a lack of scientific understanding and the authority of tradition.

NOTE Trepanation: An ancient surgical procedure involving the drilling or scraping of a hole into the skull, historically performed to treat head injuries, to relieve pressure, or for spiritual and ritualistic purposes. Evidence suggests that many patients survived the procedure.

The popularity of such methods reveals the reliance of early medicine on dogma rather than empirical evidence. In an era before the scientific method, physicians clung to established ideas, often prioritizing theoretical consistency over patient outcomes. This adherence to tradition highlights the challenges of breaking from entrenched beliefs, even in the face of repeated failure.

Although these early practices now seem archaic, they underscore an important truth: The evolution of medicine has been as much about unlearning harmful traditions as it has been about embracing new discoveries. The 19th century marked such a turning point in medicine, driven by the revolutionary insights of germ theory. Advanced by the work of Pasteur and Koch, germ theory established that microorganisms, not imbalances of humors or miasmas, were the primary cause of many diseases. Pasteur's experiments demonstrated how microbes caused fermentation and spoilage, leading him to develop pasteurization and vaccines. Koch further advanced the field by identifying specific pathogens, such as those responsible for tuberculosis and cholera, laying the groundwork for modern microbiology.

This understanding of microbes fundamentally altered the course of medicine, displacing harmful practices like bloodletting and purging. Instead of blindly attempting to "balance" the body, physicians began targeting the true sources of disease. Germ theory not only transformed individual treatments but also revolutionized public health. Techniques such as sterilization, pioneered by Joseph Lister, introduced antiseptic practices that drastically reduced infections in surgical settings. Vaccines, based on Pasteur's discoveries, offered proactive defense against deadly illnesses, marking a shift from reactive to preventive medicine.

The adoption of germ theory also catalyzed the transition from heroic medicine to a more evidence-based approach. This paradigm shift emphasized observation, experimentation, and data to guide treatments, a principle that underpins modern medical practice.

Germ theory's impact extended beyond the laboratory, driving sanitation reforms, public health campaigns, and the development of safer medical environments. It highlighted the power of science to replace superstition and tradition with rational, effective methods, paving the way for the precision medicine of today. This era was a pivotal chapter in the journey toward understanding disease at its root cause, offering hope for more targeted and successful treatments.

Antibiotics, Imaging, the Genome, and Beyond

The discovery of penicillin by Alexander Fleming in 1928 ushered in the antibiotic revolution, an enlightening era in medicine. Although its therapeutic potential was not immediately realized, subsequent efforts by Howard Florey, Ernst Chain, and others in the 1940s scaled penicillin production, making it the first widely used antibiotic. Its success in treating bacterial infections, especially during World War II, spurred the development of a multitude of antibiotics in the mid-20th century, including streptomycin, tetracycline, and erythromycin. Antibiotics fundamentally changed the treatment of infectious diseases, turning once-deadly conditions such as pneumonia, tuberculosis, and sepsis into manageable or curable illnesses. This revolution had a profound impact on global health, drastically reducing mortality rates and enabling life-saving surgeries and treatments that depended on infection control. The widespread availability of antibiotics also symbolized a paradigm shift in medicine from merely alleviating symptoms to targeting the underlying causes of disease with precision and efficacy.

However, this remarkable progress came with limitations. The overuse and misuse of antibiotics have led to the emergence of antibiotic-resistant bacteria, posing a significant threat to public health. Conditions like multidrug-resistant tuberculosis and *methicillin-resistant Staphylococcus aureus* (MRSA) highlight the urgent need for new treatments and more judicious use of existing drugs. The challenge of maintaining antibiotic efficacy underscores the delicate balance between innovation and stewardship.

The antibiotic revolution demonstrated the power of targeting the root causes of disease, laying the groundwork for the precision medicine of today. In a similar way, the introduction of X-rays by Wilhelm Röntgen in 1895 revolutionized medicine by providing a noninvasive way to peer inside the human body. For the first time, physicians could directly visualize bones, organs, and other structures, dramatically improving diagnostics and enabling more precise treatments. X-rays quickly became an indispensable tool, transforming fields like orthopedics and pulmonary medicine by allowing doctors to identify fractures, infections, and abnormalities with unprecedented accuracy.

Building on this foundation, the 20th century witnessed the evolution of medical imaging technologies. CT scans, introduced in the 1970s, combined X-ray imaging with computer processing to produce detailed cross-sectional views of the body. MRI, emerging in the 1980s, leveraged magnetic fields and radio waves to visualize soft tissues with remarkable clarity, aiding in the diagnosis of neurological, cardiovascular, and musculoskeletal conditions. Positron emission tomography (PET) scans, which detect metabolic activity, further advanced imaging by revealing functional changes in tissues, particularly in oncology and neurology.

These innovations have deepened our understanding of complex diseases like cancer. Imaging enables the precise identification of tumors, their locations, and even their metabolic behavior, informing targeted treatment strategies. This precision aligns seamlessly with the principles of modern medicine, where therapies are tailored to specific abnormalities observed in individual patients. For example, imaging plays a critical role in monitoring tumor responses to chemotherapy or guiding minimally invasive surgeries.

Medical imaging not only enhances diagnostic accuracy but also embodies the precision medicine approach by connecting observed abnormalities to personalized interventions. Coupling it with the discovery of DNA's double-helix structure created a turning point in biology and medicine. By revealing how genetic information is stored and transmitted, the integration of technologies unlocked the secrets of heredity and cellular function. This foundational breakthrough paved the way for genetic testing and the development of personalized medicine. Researchers began to understand how variations in DNA influence traits, including susceptibility to diseases, setting the stage for substantial medical innovations.

The HGP that followed was one of the most ambitious scientific endeavors of the 20th century. Its goal to map the entire human genome—over three billion DNA base pairs—provided a comprehensive blueprint of human genetic material. The project revealed the genetic underpinnings of many diseases, from cancer to rare genetic disorders, and identified countless potential drug targets. Beyond its scientific achievements, the HGP democratized genomic data, making it widely accessible to researchers and clinicians worldwide.

Subsequent application of genomics in medicine has changed the healthcare landscape. Precision oncology, for example, uses genomic profiling to identify mutations driving a patient's cancer, enabling targeted therapies that attack tumors while sparing healthy cells. Gene therapy has also emerged as a powerful tool for correcting genetic disorders by repairing or replacing faulty genes, offering hope for conditions like spinal muscular atrophy and hemophilia. Additionally, genomics has revolutionized the diagnosis and management of rare diseases, many of which were previously undiagnosed or misunderstood. Whole-genome sequencing now allows clinicians to identify the root causes of these conditions, facilitating earlier interventions and tailored treatments.

From Blockbuster Drugs to Targeted Therapies

The history of modern medicine also exhibits a gradual shift from broad-spectrum medications, such as aspirin, to highly targeted therapies designed to address specific biological mechanisms. Blockbuster drugs, widely prescribed for a broad range of patients, once dominated the pharmaceutical landscape. Although effective for many, these drugs often lacked precision, leading to variable outcomes and unintended side effects. Advances in molecular biology and genomics have ushered in a new era of targeted therapies, including biologics and monoclonal antibodies, that focus on individual genetic and molecular profiles.

Targeted therapies, such as trastuzumab for HER2-positive breast cancer patients, have had a revolutionary impact on cancer treatment. Immunotherapies, such as checkpoint inhibitors, have also emerged as powerful tools, leveraging the patient's immune system to recognize and destroy cancer cells.

NOTE   A checkpoint inhibitor is a type of immunotherapy drug that helps the immune system recognize and attack cancer cells. These drugs work by blocking immune checkpoints—molecules that act as "brakes" on immune responses—allowing T cells to become more active in fighting cancer.

These therapies mark a significant departure from traditional chemotherapy, offering improved efficacy and fewer side effects for many patients.

The benefits of tailoring treatments like these to individual profiles are profound. Precision approaches can maximize therapeutic effectiveness, minimize adverse effects, and improve overall patient outcomes. However, these advancements come with significant challenges. Targeted therapies are often expensive, limiting accessibility for many patients. Scaling precision approaches to broader populations is further complicated by the complexity of developing and manufacturing such therapies. Additionally, healthcare systems face logistical hurdles in integrating genetic testing and personalized treatment plans into standard practice.

Despite these challenges, the shift from broad-spectrum medications to targeted therapies represents a critical step in the evolution of medicine. By addressing the unique needs of each patient, these innovations embody the promise of precision medicine, redefining healthcare for a new generation.

Practice to Precision: Lessons Learned

The transition from rudimentary practices to evidence-based approaches underscores the iterative nature of medical progress: each breakthrough builds on the knowledge—and the missteps—of the past. This historical journey highlights the critical importance of precision. Broad, one-size-fits-all approaches, although helpful in their time, often fell short of addressing the complexities of individual patients. Precision medicine builds on this history by integrating genomics, digital health technologies, and advanced analytics to tailor care to the individual. This approach not only improves outcomes but also exemplifies medicine's ultimate goal: to treat each patient with the specificity and care they deserve.

Reflecting on the journey reveals an essential truth: progress in medicine is a continuous process fueled by curiosity, perseverance, and a willingness to adapt. The path from bloodletting to precision medicine outlines a profound evolution in healthcare. What began as harmful, generalized practices based on limited understanding has progressed to leading-edge therapies tailored to the unique needs of each individual. Advances at each step have not only reshaped the way we diagnose and treat diseases but also redefined our understanding of health itself. Precision medicine, with its focus on individualized care, represents the culmination of centuries of innovation, offering treatments that are more effective, less invasive, and deeply personal.

Yet this progress has not been without challenges. Emerging threats like antibiotic resistance, disparities in access to care, and the complexities of scaling personalized treatments underscore the need for continual innovation. As history has shown, the path forward requires persistence, creativity, and a commitment to learning from both successes and failures.

Looking ahead, precision medicine holds the promise to revolutionize healthcare for the next generation. By integrating advances in genomics, AI, and real-time health monitoring, it has the potential to prevent diseases before they occur, optimize treatments for existing conditions, and empower individuals to take a more active role in their health. Precision medicine is not just the future of healthcare—it is a testament to how far we have come and a beacon of hope for what lies ahead.

Integrating the Outcomes: The Rise of Individualized Care

Individualized care represents the culmination of two evolving processes: the ability to gather personalized data and the capacity to integrate this information into actionable insights. Advances in wearables, genomics, and biomarkers have made it possible to collect vast amounts of data tailored to each individual, from real-time health metrics to comprehensive genetic profiles. Yet the true power of individualized care lies not only in data collection but also in synthesizing this information through technologies like Big Data analytics and AI. Together, these tools enable healthcare providers to move beyond generalizations, crafting precise and highly effective solutions for each patient.

NOTE Assembling information about humans as individuals and integrating it with knowledge about humans collectively is the essence of precision medicine.

By tracing the evolution of medical innovation, this chapter provides the foundation for how the integration of individualized information is shaping today's healthcare landscape. From early attempts at personalization to leading-edge therapies, examination reveals how the interplay of data and integration is central to achieving the promise of precision care. Understanding this evolution offers a deeper appreciation of the complexities of modern medicine and the innovative approaches yet to come.

Early Efforts to Personalize Medicine

As we have seen, attempts to tailor treatments to individual patients are not a modern phenomenon. TCM emphasized the balance of yin and yang and the harmony of internal energies, tailoring treatments to a person's constitution and symptoms. Similarly, humoral theory, rooted in ancient Greek medicine, sought to balance the four bodily humors through practices like bloodletting, purging, and dietary adjustments. Both approaches acknowledged that patients are unique and require personalized interventions, laying the groundwork for the broader concept of individualized care.

However, these early efforts were limited by the tools and knowledge available. Observational methods, although insightful for their time, were subjective and lacked the precision needed to identify the biological mechanisms underlying diseases. TCM relied on practitioner intuition and symptom patterns, and humoral theory was based on a framework of balance without a clear understanding of anatomy, physiology, or disease pathology. Treatments were often guided by tradition rather than empirical evidence, leading to variable and sometimes harmful outcomes.

The primary constraint on early personalized care was the inability to gather and synthesize complex data. Without tools to measure, analyze, or integrate physiological, genetic, or environmental factors, these systems could not achieve the level of precision seen today. Despite their limitations, they introduced the fundamental idea that healthcare should address the individual rather than the disease in isolation. These early efforts highlight the longstanding desire to personalize medicine, even in the absence of advanced tools. Diagnostics built on these efforts, evolving from basic symptom observation to advanced tools that provide detailed insights into the human body. In ancient times, healers relied on visible signs, such as skin changes or fever, and subjective symptoms reported by patients. Although helpful, these methods were limited in precision and often led to generalized treatments.

The advent of laboratory testing and imaging technologies revolutionized diagnostics in the 19th and 20th centuries. Blood tests provided objective measurements of organ function, hormone levels, and markers of disease, enabling clinicians to identify conditions with greater accuracy. Imaging tools like X-rays, MRIs, and CT scans allowed physicians to visualize internal structures noninvasively, transforming diagnosis and treatment planning. These advancements laid the groundwork for more personalized approaches by linking specific diagnostic findings to targeted interventions.

In the modern era, wearable devices have added to this, providing real-time data on metrics such as heart rate, sleep patterns, and blood sugar levels. Unlike traditional diagnostic tools used intermittently in clinical settings, wearables generate continuous streams of personalized data, empowering individuals to monitor their health proactively while giving clinicians valuable insights into their patients' daily lives.

This collection of diagnostic tools serves as the foundation of individualized care by providing the raw data needed for precision interventions. They generate a wealth of information that, when integrated with other datasets such as genomic or environmental data, creates a holistic view of each patient. This integration exemplifies how modern diagnostics not only identify diseases but also drive the personalized solutions that define precision medicine.

Genomics: The Expansive Blueprint of Individualized Care

The rise of genomics was the next step in this evolution, unveiling the genetic blueprint that underpins human health and disease. With the decoding of the entire human genome in 2003, the medical community was provided with an unprecedented view of the genetic variations that influence health and underpin individualized medicine. Genomics now plays a central role in predicting disease, optimizing therapies, and developing precision drugs. By identifying genetic markers associated with specific diseases, genomics enables early detection and preventive strategies tailored to at-risk individuals. Pharmacogenomics, as we have seen, ensures that medications are chosen and dosed based on a patient's genetic profile, reducing adverse effects and improving efficacy. Moreover, the data generated by genomic sequencing is vast and complex, presenting both opportunities and challenges. Each genome contains millions of data points that can be analyzed, interpreted, and integrated with clinical information to derive actionable insights. Advanced analytics, including AI and machine learning, are increasingly essential for processing this wealth of information. These tools can identify patterns, predict risks, and suggest interventions, making genomics a cornerstone of precision medicine.

The rise of diagnostics, wearable devices, and genomics and the corresponding massive amounts of data have created new challenges. Wearable devices continuously collect health metrics such as heart rate, activity levels, and glucose levels, and genomic sequencing produces millions of data points for each individual. These streams of data are not only vast in volume but also highly complex and heterogeneous, encompassing genetic, physiological, behavioral, and environmental factors.

Traditional methods of data analysis struggle to keep pace with this influx. Healthcare systems and clinicians often lack the tools to synthesize these diverse datasets into actionable insights. For example, although a fitness tracker may provide continuous heart rate monitoring and a genomic profile may reveal disease predispositions, integrating these disparate data streams to inform a personalized treatment plan remains a significant hurdle. The complexity of correlating real-time wearable data with genomic information, clinical histories, and environmental influences has outstripped the capabilities of conventional analytical approaches.

This challenge explains the role of Big Data analytics and AI in healthcare. Big Data technologies can handle the volume and complexity of these datasets, enabling the integration of diverse sources into cohesive, patient-specific profiles. AI further enhances this process by identifying patterns, predicting risks, and generating recommendations tailored to the individual. Machine learning algorithms, for instance, can detect subtle correlations between wearable data and genetic markers, providing early warnings or guiding interventions.

Data Gathering to Integration: The Big Picture

To be successful, the evolution of individualized care hinges on a pivotal shift: moving from data gathering to integration. Advancements in genomics and wearable technology have enabled the collection of vast amounts of personalized data, but their true potential is realized only through integration. Big Data analytics and AI are critical for transforming raw, fragmented data into precise, actionable insights.

Integration occurs when diverse datasets—such as genomic profiles, real-time metrics from wearables, and clinical histories—are combined and analyzed to inform personalized care. In oncology, for example, AI-driven predictive models use genomic data to identify mutations driving a patient's cancer while integrating real-time health metrics, such as physical activity or treatment side effects, from wearables. This holistic view allows clinicians to refine treatment plans dynamically, tailoring therapies like immunotherapy or targeted drugs to the individual's unique needs.

The power of integration lies in its ability to synthesize complexity. Big Data analytics aggregate vast datasets, identifying patterns and relationships that would be impossible to detect manually. AI then takes this process further, learning from these data to predict outcomes, flag potential risks, and recommend interventions. For instance, predictive models can estimate how a patient with a specific genetic mutation might respond to a treatment, enabling clinicians to choose the most effective option with confidence. These insights can also guide preventive care, using real-time data to intervene before complications arise.

For patients, the benefits are equally significant. Integrated data systems provide individuals with personalized dashboards, offering clear insights into their health and empowering them to make informed decisions. Real-time alerts, such as warnings of irregular heart rhythms detected by wearables, ensure timely action, enhancing outcomes and safety. This integration marks a new era in healthcare, where data gathering is no longer the end goal but the first step toward precise, individualized solutions. By transforming complexity into clarity, Big Data and AI enable clinicians and patients to navigate healthcare with unprecedented precision and confidence.

This integration of vast amounts of personal health data into actionable healthcare solutions brings with it significant ethical and practical challenges, though. As genomic profiles, wearable metrics, and clinical data are synthesized, concerns about privacy, ownership, and equity take center stage. Sensitive data, such as genetic predispositions and real-time health metrics, could be vulnerable to misuse if not properly safeguarded. Questions about who owns the data—patients, healthcare providers, or technology companies—complicate its ethical use, especially in a landscape increasingly dominated by commercial interests. Additionally, disparities in access to integrated healthcare systems raise concerns about equity, as underserved communities risk missing the benefits of precision medicine.

Practical challenges also hinder the widespread implementation of integrated systems. Developing the infrastructure to collect, store, and process vast datasets requires significant investment, making it challenging for many healthcare systems, particularly in resource-limited settings. Training clinicians and technicians to utilize Big Data analytics and AI effectively adds another layer of complexity. Moreover, the cost of genomic sequencing, advanced wearables, and AI-driven platforms often limits access to wealthier patients and institutions, aggravating existing healthcare disparities.

To ensure that the benefits of integration are accessible broadly, a commitment to responsible implementation is essential. Robust data privacy regulations must be enforced to protect sensitive information, and policies addressing ownership ensure that patients retain control over their health data. Efforts to subsidize infrastructure costs, expand training programs, and democratize access to advanced tools are equally critical. Equitable integration requires not only technological innovation but also systemic reforms that prioritize inclusivity, and we will visit these specific imperatives in Chapter 9.

Integration as the Future of Medicine

The future of medicine lies in the reliable integration of data to deliver increasingly individualized care while simultaneously ensuring necessary access, privacy, and safeguards in its deployment. Emerging trends, such as the use of AI in predictive analytics, are already reshaping healthcare. Future advancements in data integration promise even greater precision. Combining genomic information with real-time data from wearables and environmental factors, such as air quality and lifestyle metrics, will create a holistic view of each patient's health. This interconnected approach will enable physicians to understand how genetics interact with daily behaviors and surroundings, allowing for hyper-personalized prevention strategies and treatments.

The evolution of medical innovation has brought healthcare to a pivotal moment, where the integration of personalized data streams unlocks the full potential of precision medicine. From the early days of basic symptom observation to the advanced technologies of genomics, wearables, and Big Data analytics, each breakthrough has built toward the ability to gather and synthesize individual health information. This integration marks the culmination of centuries of progress, transforming raw data into actionable solutions that address each patient's unique needs.

This chapter bridges the gap between data collection and actionable healthcare solutions, highlighting how integration drives modern precision care. The tools explored—such as wearable devices, genomic sequencing, and AI-driven analytics—illustrate how diverse data streams are combined to provide a holistic view of health. By leveraging this interconnected approach, clinicians can predict risks, personalize treatments, and deliver real-time insights that empower patients to take control of their well-being.

Looking ahead, the revolutionary potential of integration is limitless. As technologies continue to evolve, healthcare will become increasingly adaptive, predictive, and proactive. The integration of personalized data streams will redefine the relationship between patients and providers, creating a system that prioritizes prevention, precision, and individual empowerment. This new era of healthcare is not just about treating diseases but about understanding and optimizing health in ways that were once unimaginable.

Integration unlocks the promise of precision medicine, offering a future where care is more effective, equitable, and empowering than ever before; as precision medicine continues to evolve, its ability to tackle some of healthcare's most stubborn challenges becomes increasingly clear. We've seen how personalized interventions transformed Michael and Jennifer's weight-loss journey, and how Emily found clarity through pharmacogenomics. Yet for patients living with chronic pain, the path to relief remains one of the most complex and frustrating. Too often, treatments are prescribed through trial and error, leaving individuals trapped in a cycle of ineffective medications, unwanted side effects, and diminishing hope.

Now we make a pivotal transition in our examination of healthcare. In previous chapters, we explored how tools like wearables and genomic sequencing revolutionized data collection, emphasizing the importance of understanding individual variability. In the immediately preceding sections we have built the bridge to the second half of the book, where our focus shifts to the integration of these datasets: combining them with clinical expertise and advanced technologies to deliver actionable, precise interventions that define modern precision medicine.

For Amy, whom we meet in Chapter 7, this struggle is all too familiar. Once an active and energetic professional, she now finds herself constantly negotiating with pain—adjusting her life around it, rather than overcoming it. Each new prescription brings a mix of anticipation and disappointment, as doctors seek solutions without a clear roadmap. But what if the answers have been there all along—hidden within vast datasets, waiting to be uncovered? In the next chapter, we explore how the power of Big Data analytics helps rewrite Amy's story, turning a pattern of failed treatments into a breakthrough in personalized pain management.

Notes

1. Hippocrates of Cos, *The Oath*, trans. Jones, William, Loeb Classical Library 147 (1923): 298–299.

2. Hildegard of Bingen, *Hildegard's Healing Plants: From Her Medieval Classic Physica*, trans. Hozeski, Bruce, Beacon Press (2002).

3. Avicenna, *The Canon of Medicine*, trans. Bakhtiar, Laleh, Kazi Publications (1999).

4. Aberth, John, *The Black Death: A New History of the Great Mortality in Europe, 1347–1500*, Oxford University Press (2005).

   Broadberry, Stephen, Campbell, Bruce, and van Leeuwen, Bas, "English Medieval Population: Reconciling Time Series and Cross-Sectional Evidence," London School of *Economics*, March 3 (2011).

5. Ibid.

6. Vesalius, Andreas, *On the Fabric of the Human Body, trans.* Richardson, William, and Carman, John, Norman Publishing (1998).

7. Flexner, Abraham, *Medical Education in the United States and Canada: A Report to the Carnegie Foundation for the Advancement of Teaching*, Carnegie Foundation (1910).

8. Dattani, Saloni, Spooner, Fiona, Ritchie, Hannah, and Roser, Max, "Antibiotics and Antibiotic Resistance" (2024), `https://ourworld indata.org/antibiotics` (licensed under CC-BY).

9. Vonderlehr, Eugene, Wenger, Oliver, and Heller, John, "Untreated Syphilis in the Male Negro," *Journal of the American Medical Association* 107, no. 11 (1936): 856–860.

   Rockwell, Donald, Yobs, Anne, and Moore, Brittain, "The Tuskegee Study of Untreated Syphilis: The 30th Year of Observation," *Archives of Internal Medicine* 114, no. 6 (1964): 792–798.

10. Rye, Connie, Wise, Robert, Jurukovski, Vladimir, DeSaix, Jean, Choi, Jung, and Avissar, Yael, "14.2 DNA Structure and Sequencing," (2022), `https://openstax.org/books/biology/pages/14-2-dna-structure-and-sequencing` (licensed under CC-BY).

The Joy and Pain of Being Amy

Amy pulled her blouse from the dryer, pressing it to her face. The scent of detergent barely masked the familiar smell of food that clung to the fabric, no matter how many times she washed it. She didn't mind. It was the smell of work—of movement, purpose, and life that, despite everything, kept going.

She smoothed it with care, removing the wrinkles like a ritual, her hands moving automatically through the motions. Washing, straightening, preparing—not just her outfit, but herself. Arriving at the Metro stop, she shifted her weight, rolling her shoulders, gauging the deep, dull ache in her lower back. A morning assessment: not good, not terrible. Manageable.

When Amy finally stepped into the restaurant of the downtown office building, the air smelling of sanitizer, breakfast, and coffee, she took a deep breath. The sounds of clattering utensils and murmuring employees wrapped around her. This place had its own rhythm—one she knew by heart. She exhaled and stepped forward, ready for another day.

The Value of Work

Amy had been working since she was 16. First a summer job at the mall, then the diner, then anywhere that paid a little more money. Now, here—decades in, it felt even longer. She didn't just do the job—she was the job. She knew the regular customers before they ever placed their order: Mr. Calloway, coffee black, no sugar; Denise from the law office, bagel with cream cheese, double latte, napkin folded neatly in the bag. She knew who would tap their fingers impatiently, who would crack a joke, who had been coming in since before she started and still didn't know her name. The rhythm, the predictability—it steadied her, gave her something to hold onto, especially on the busiest days. But routine had a way of making her invisible, too. Customers saw the food served, confirmed with a nod, but didn't see the person behind it. Some days, Amy didn't mind. Other days, it pressed as heavily on her as the ache in her back. But she showed up and did the work, and that was often enough.

Amy tidied the hostess stand as Teri, one of the new hires, groaned. "God, my feet are killing me," the girl whined, rolling her ankles. Amy suppressed a laugh, "You don't know pain yet, sweetheart." The younger ones always complained about sore feet, long shifts, or rude customers like they were facing a war zone. Amy had been through it all—double shifts that stretched into all day, customers who snapped instead of saying her name, paychecks that were too little, and rent that refused to be late. Pain wasn't just in her feet. It lived in her back, her shoulders, her knees, a dull, constant ache that she carried like a second skin. But she didn't stop. She couldn't.

To customers, she was just Amy—smiling, handling orders, moving fast so they didn't have to wait. To her boss, she was reliable, always saying yes when others called off. To her family, she was the one who was always there, always fine. To herself? She was tired, soul-tired, the kind of tired that sleep didn't fix. But the next customer arrived, and Amy, like always, kept moving.

Amy was completing an order when she saw Emily approaching, her usual warm, enthusiastic presence cutting through the midday rush. A go-getter from one of the firms upstairs, Emily came in at least three times a week, always greeting Amy like an old friend. As Amy handled the order, Emily studied her. "You're always smiling, Amy. How do you do it?"

Amy chuckled, tucking a stray strand of hair behind her ear. "If I stopped, you'd know something was wrong, wouldn't you?" Emily laughed, shaking her head. "Guess that's true. Some days I can barely manage a smile, and I sit in an office most of the day." She added a tip to the receipt and winked. "You're a better woman than me."

Amy smiled again—because that's what she did. But as Emily turned to leave, she felt the familiar tightness in her lower back, the lingering soreness in her fingers from handling things all day. Emily didn't realize Amy had told the truth. She smiled because it was expected. Because if she stopped—if she let the exhaustion, the ache, the sheer weight of it all show—people would notice. And noticing led to questions Amy didn't want to answer.

The next customer arrived, and without thinking, she smiled again.

The Physical Toll

Pain had become as much a part of Amy's routine as beginning the day. It was there when she woke up, a dull throb in her knees before she lowered her legs from the side of the bed. It settled into her lower back by the time she stepped onto the Metro, stiffened her hands as she readied for work. By the middle of her day, it pulsed in her ankles, crept up her spine, an unspoken reminder that her body was keeping score.

She had learned to move through it, to ignore the way her muscles protested each step. Her job didn't wait for pain. Customers kept arriving, orders kept coming, and the work kept going.

Teri, the younger coworker, stretched dramatically, groaning about needing a break. Amy just smiled wryly and shook her head. Breaks were for the young. When she was their age, maybe she would've taken five minutes, rolled out her shoulders, and sat for a moment. But now? Stopping made it worse. Slowing down let the aches settle in deeper, made it harder to start again. So she kept moving. Fresh iced tea needed brewing. She filled the container, her fingers stiff as she replaced the top. The kitchen clattered as she stood nearby, the warm air prickling against her blouse. Every motion had a cost, but she paid it without complaint. Because that's what work was—doing what needed to be done. Amy had work to do.

The early dinner crowd was beginning to arrive when Amy heard her name. "Hey, Amy—can you stay late this evening?" Her boss barely looked back from the kitchen, already knowing the answer. She hesitated, just for a second; she hadn't sat down in hours. Her back was stiff, her hands sore, her head pounding. But she knew why he was asking. "Yeah, I got it," she said, forcing a smile. "Appreciate it." And just like that, he was gone, onto the next thing, never questioning if she'd actually wanted to stay. The younger employees called off without a second thought, knowing someone else would cover. She never made a fuss about it, never called off unless she was practically dying. She wasn't the loudest or the fastest, but she was dependable. And in a job like this, that was half the battle.

She wiped her hands and stepped to the front of the restaurant. Another wave of customers would come in soon. The job wasn't glamorous, but it was necessary. That was the hidden cost of being dependable: people leaned on you. They built their schedules around your reliability, and when you were always there, always working, always saying yes, they stopped noticing that you might need a break, too. Amy straightened, smiled, and got back to work.

Amy's Younger Years: A Different Kind of Hard Work

Amy used to have energy, real energy, the kind that made her feet light, let her dance in the kitchen while cooking dinner, and carried her through long shifts without a second thought.

She married young—too young, maybe. Two kids before she turned 25, and a husband who was always on the go, chasing the next job. Stability wasn't easy, so she became the constant, the one who made sure the fridge stayed full and the kids had what they needed.

She started in retail: folding shirts, ringing up customers, organizing displays that no one would keep tidy for more than five minutes. It was exhausting, but a different kind of exhausting. Back then, she could stay on her feet all day and still have something left for herself at night. She danced then, too—actually danced. In the living room, in the kitchen, in the aisles of the grocery store with her kids giggling at her. She moved freely, without thinking about how much it would hurt later. But retail jobs paid little and offered even less—no security, no benefits, you name it—so she switched to restaurants, where the paychecks were steadier, the tips unpredictable, and the work relentless.

The pain started small. A little stiffness in her knees after a long shift, a dull ache in her lower back that lingered but always faded by morning. Nothing she couldn't handle, so she ignored it. What else was she supposed to do? Everyone in this line of work had aches and pains—it was part of the job, part of life. Then it got worse. The ache in her knees didn't go away overnight. Her back hurt before she got out of bed. Her wrists started throbbing after hours of handling the register, gripping heavy bags, wiping down counters. By the end of some shifts, her ankles felt like they were burning, every step sending sharp little reminders through her body.

Pain was something to push through, not something to stop for. Stopping meant losing hours, losing money. She told herself it would pass, but it didn't. One morning, she bent to slide on her shoe and felt something in her back seize. The sharp, breath-stealing kind of pain. She stayed like that for a moment, hands braced on her knees, waiting for it to pass. It did, eventually, but it left something behind—a new awareness. She wasn't just sore, wasn't just tired. This pain wasn't going anywhere; it had settled into her, become part of her. And no matter how much she tried to ignore it, she knew it wasn't done with her yet.

Amy didn't go to the doctor right away. She told herself it wasn't that bad, that she could handle it; but one morning, she could barely get out of bed. The pain in her back was sharp, unforgiving, and her knees ached so badly that the simple act of standing felt like punishment. She finally made an appointment. The doctor listened, nodding as she described the stiffness, the soreness, the way her body felt like it was giving up on her. Then he turned, "Wear and tear," he said. "Part of aging." That was it? "Nothing to do but manage it," he said. Stretching, ice packs, over-the-counter medications that barely touched the pain. Rest, if she could, but resting meant losing hours, and losing hours meant losing pay.

She tried the stretches—when she remembered. She took the medications, but they dulled the edges only slightly; the pain was still there, waiting. A specialist might have more answers, but she didn't have time for specialists and didn't have the money for appointments that led nowhere. So she did what she'd always done—she worked through it. She showed up, smiled, moved through the pain. Because that was the only option she had.

A Conversation with Mark

Amy's husband, Mark, watched as she lowered herself onto the couch one evening, slowly and carefully, pressing a hand to her lower back. She tried not to wince, but he saw it anyway. "Maybe you should think about leaving the job," he said quietly.

Amy let out a dry laugh. "And do what? This is what I know." Mark sighed, rubbing the back of his neck. "I don't know. We'll figure it out." They never did.

Not because they didn't want to. Because figuring it out meant options— options Amy didn't have. It meant walking away from steady pay, even if the hours were long and the work unforgiving. It meant retraining, going back to school, or starting over in something else. But who hires someone with pain and no degree? So they never talked about it past that. Instead, Mark did what he could. Massaged her shoulders at night. Rubbed her feet even though she swore they were too rough. Made her favorite tea when she had a particularly grueling shift. It was love in its simplest form—not grand gestures, but the small, simple ways he tried to lighten the weight she carried. And Amy? She smiled, leaned against him, and pretended it was enough.

Amy worked because she had to. Because bills didn't pay themselves, because groceries didn't appear in the fridge by magic, because her family needed her to. She worked because that's what she'd always done. Since she was young, punching the clock, feeling the pride of that paycheck in her hands. She learned early that work meant security, that tired feet at the end of the day meant food on the table, that showing up, even when it was hard, mattered.

So she kept going. Even when her body protested and the pain settled into her bones like an unwelcome houseguest. Even when Mark, with that familiar mix of love and exasperation, told her to slow down. "What, and miss out on all the fun?" she'd tease, pouring herself into the chair. Mark would just shake his head. "You're impossible." And maybe she was. Maybe she didn't know how to stop. Maybe, even if she could, she wouldn't. Because deep down, she was still that girl who found comfort in routine, in movement, in the simple act of doing. So she stretched, rolled her shoulders, and got ready for another day. Because work was work—and also who she was.

A Lifetime of Shrugged Shoulders

Amy flipped through an old magazine, barely seeing the pages. The waiting room smelled of antiseptic and worn carpet, a scent she knew well. A few other patients sat scattered around, some staring at their phones, others just waiting, like her. She already knew how this visit would go. The nurse would call her back, check her weight, take her blood pressure. Ask what brought her in today, although the answer was the same as last time—and then the doctor would come in. Barely glance at her chart; barely look at her. "It's just part of getting older."

Amy had lost count of how many times she had heard that phrase. It was the familiar refrain, the same script, different office. They would suggest the usual—stretching, a referral for physical therapy, maybe some topical balm. No one ever asked about her job. About how many hours she spent on her feet, lifting, bending, moving. About the exhaustion that made her bones feel heavier than they should. If they did, maybe they'd understand why physical therapy wasn't practical. Why "just resting" wasn't an option. She wanted to ask—what happens when the pain doesn't go away? When it's more than just "getting older"? But she already knew how they would respond.

Amy sighed, closing the magazine. The nurse called her name, and she stood, bracing against the dull ache in her knees. Another visit, another shrug.

The False Hope of Painkillers

It took years, but eventually, a doctor wrote her a prescription. "These should help," he said, handing her the slip. "Take them as needed."

Amy hesitated. She didn't like taking pills, but she was desperate. The pain was wearing her down, turning her days into marathons she barely finished. If this was what it took to keep going, maybe it was worth a try.

The first pill made her feel . . . off. The pain dulled, but so did everything else. She felt foggy, like her thoughts were wading through syrup. Her stomach churned, a deep nausea that lingered through her day. She moved slower. Forgot things. Had to double-check things, and sometimes triple-check. Her focus, once sharp despite her exhaustion, felt like it was slipping. By the end of the first week, she realized something terrifying: work wasn't easier. It was harder. She pushed through, hoping her body would adjust. But the pills didn't just take the edge off the pain. They took the edge off her. And then there was the fear.

She had heard the stories—people who started with just a few pills, taking them only when they needed to, until one day, they needed them all the time. Until the pills took over. She wouldn't let that happen. So one morning, she simply stopped.

The pain came back—like an old enemy waiting in the shadows—but at least she felt like herself again. At least the pain was hers. Hers to carry, hers to manage, hers to push through. And like always, Amy kept moving.

It had been a brutal afternoon. The kind that left her body feeling like it had been wrung out and tossed aside. By the time she made it home, she could barely walk. Her knees were swollen, stiff, and aching with every step. Her lower back burned, the pain radiating deep, sharp, unrelenting.

Mark took one look at her and shook his head. "Amy, you need to go in." "I'll be fine," she muttered, wincing as she eased onto the couch. "No, you won't." He crouched in front of her, resting a hand on her knee. "Come on. Just this once, let's not argue about it."

She wanted to fight him on it. Wanted to tell him that it was pointless, that nothing ever changed. But the pain was different now—worse. And she didn't have the energy to argue. She let him help her to the car.

The Feeling of Futility

The hospital emergency room was packed, people slumped in chairs, shifting uncomfortably, murmuring in low voices. Amy took a seat and settled in to wait. Hours passed as she watched the clock, listened to the nurses call names, watched people with more "urgent" problems taken ahead of her. By the time they finally called her, it was only a few hours from dawn.

She sat stiffly on the paper-covered exam table, exhaustion and pain pressing down on her like a weight. The doctor, young and hurried, flipped through her chart for several seconds before looking at her. "Have you considered weight loss?" Amy blinked. For a moment, she thought she must have misheard. That's what he had to say? After everything? After years of working through pain, of doctors shrugging her off, of trying every worthless suggestion thrown her way?

She wanted to scream. I have considered everything. I have stretched, I have iced, I have swallowed pills that made me feel like a stranger in my own skin. I have worked through pain that would break other people. And you think the answer is weight loss? She just nodded. Because what else was there to say? The doctor scribbled something in her chart and moved on, already thinking about the next patient. And Amy sat there, the pain still burning, knowing nothing had changed.

Amy stepped out of the ER into the early morning air, the sky streaked with the last traces of night. The city was waking up—cars humming down the streets, people going to their jobs, the world moving like nothing had happened. She moved slowly, each step a reminder of why she had come in the first place. The weight of her body, her pain, her exhaustion—she carried it alone.

Mark opened the passenger door, watching her carefully, waiting for her to say something. She had nothing to say. Nothing had changed; no solutions, no relief, just another doctor, another shrug, another dismissal wrapped in clinical politeness. Have you considered weight loss? That was all he had offered her.

Amy let out a breath and sank into the seat, pressing her head against the window. She wondered how many other people—people like her—had walked out of offices like this, carrying the same burden. How many had sat in waiting rooms for hours, only to be told something they had already heard, something useless? How many had hoped, even just a little, that this time would be different? She closed her eyes. The pain and exhaustion hadn't changed. But at least home was waiting, a bed, a few hours of sleep, and then another shift. Because the world didn't stop for pain, and neither could she.

A Voice of Concern

The coffee machine hissed as Amy helped the kitchen staff, the scent curling into the air. She moved through the motions like she always did, with a smile. Another day, another customer, another ache. Her joints were particularly bad today, stiff and sore from the constant repetition of turning, lifting, moving. Without thinking, she rubbed her hands as she helped another customer, flexing her fingers to shake out the discomfort.

That's when she noticed him. A man in a crisp button-down and glasses stood just inside the restaurant, watching her hands. Not impatiently, not like a customer waiting too long for their turn—just . . . observing. When she handed him his change, he spoke.

"Chronic pain?" Amy let out a dry, tired laugh. "Is there any other kind?" She expected the usual response—a polite chuckle, a nod, maybe some vague comment about how tough that must be. People always meant well, but small talk didn't help. Instead, his next words stopped her cold.

"There might be something that can actually help."

Amy blinked. Help? Real help? Not just ice packs, balms, medications, and useless advice? She wasn't sure whether to scoff or listen. But something in his voice—calm, certain—made her pause. For once, someone wasn't just shrugging their shoulders. And that was enough to make her curious. Amy handled the order for the next customer, still rubbing absent circles into her wrist. The man in the button-down hadn't left yet. He studied her for another beat before speaking again.

"Have you ever had your pain analyzed using precision medicine?" Amy blinked. "My what now?"

She expected him to say something about stretching or better shoes, maybe even suggest she "take it easy." But no one had ever asked her to think about her pain in a way that wasn't just enduring it.

He smiled, adjusting his glasses. "I work at the Medical Center. We're running a program looking at chronic pain differently, not as something to just 'push through,' but as something to understand." Amy arched an eyebrow. "Understand?"

"Instead of treating pain like a generic condition, we look at you. Your unique biology, your job, your daily movements, your genetics—even environmental factors. We don't guess what might help: we use real data to figure out what will." Amy stared at him, skeptical but intrigued.

"What's the catch?" she asked, folding her arms. "No catch," he said. "Just the chance to figure out what's happening to you—and do something about it."

Amy wasn't sure she believed him. But it was the first time someone had treated her pain as something that mattered enough to study.

Treating the Person, Not the Symptoms

Amy leaned against the counter, arms crossed, skeptical but listening. The man—Dr. Ben Miller, he had introduced himself—watched her reaction carefully before continuing.

"Most pain treatments today are a guessing game," he said. "Doctors throw the same pills at everyone, the same physical therapy, the same generic advice. But pain isn't one thing. It's different for every person. So why should we treat it like it's all the same?" Amy shrugged, "You're asking the wrong person. I just do what they tell me." Dr. Miller shook his head. "And that's the problem. We've been telling you the wrong things." She raised an eyebrow. "Oh yeah?"

He nodded. "We use Big Data and advanced analytics at the Medical Center to study chronic pain—not just as a symptom, but as a condition with a cause. We analyze thousands of cases, looking for patterns, and we've found that different people respond best to different interventions. Some need targeted physical therapy that actually considers their work and daily movements, and some need nerve stimulation to calm overactive pain signals. Others need non-opioid medications tailored to their genetic profile."

Amy frowned. "Genetic profile?" "Yes," he said. "Your genes affect how you process pain—whether your nerves overreact, whether inflammation lingers, even how well certain medications work for you. Instead of guessing, we can now test. We can see what's happening in your body specifically and target treatment based on real data." Amy glanced down at her hands, flexing her aching fingers. "So . . . instead of just masking pain—" "We figure out why it's happening in the first place," Dr. Miller finished. Amy exhaled slowly.

For years, doctors had treated her pain like an inconvenience—something to manage, to endure, to accept. No one had ever suggested it was something with a cause. She wasn't sure she trusted the new approach yet, but now, someone was telling her that her pain mattered. Dr. Miller didn't push. He just watched her, patient, steady, like he knew exactly what she was thinking. And maybe he did—because unlike the doctors who dismissed her, he wasn't handing her a prescription, wasn't telling her this would "fix" her. He wasn't even promising a cure . . . he was just seeing her.

"Would you be open to learning more?" Amy hesitated. She could hear Mark's voice in the back of her head—"Take it easy." But something about this felt different. She exhaled, rolling her sore arms as if weighing the decision.

Finally, she nodded. "I don't have anything to lose." Dr. Miller smiled. "That's what I was hoping you'd say." Amy didn't know if this would lead anywhere. But after years of empty advice, she wasn't just being told to endure.

The Power of Knowing Why

Amy stood outside the Medical Center, staring at the glass doors. She had almost talked herself out of coming—twice. Mark had asked if she was sure about this. She wasn't, but here she was, already sore and exhausted from another morning, walking into a place that felt worlds away from the crowded waiting rooms she was used to. Inside, the air was cooler, quieter. No rushed doctors, no ringing phones, no endless rows of sick people flipping through months-old magazines.

A researcher greeted her—not a nurse, not a receptionist, but someone who actually seemed eager to explain what was about to happen. "We're not here to guess," the woman said. "We're here to measure." They walked her through the process: DNA testing to see how her body metabolized medication—whether the pain medications she had tried were the right ones for her; environmental mapping to check for long-term exposure risks—events or activities in her daily life that could be making her pain worse; and wearable technology to track her pain patterns in real time, mapping when and why her symptoms flared.

It was strange—almost too good to be true. For years, doctors had treated her pain like an abstract concept, something to be managed with a shrug and a pill bottle. Now they were treating it like something tangible, something that could be studied, something that might actually have answers. Amy exhaled slowly, flexing her aching fingers as she listened. Could this really tell her something?

The Science of Pain

Ten days later, Amy sat in the consultation room, her fingers tapping restlessly against the arm of the chair. She wasn't sure why she was so nervous. Maybe because this was different. Now there was a chance she might get real answers instead of another vague recommendation to stretch more or just "listen to her body." The door opened, and Dr. Miller walked in, tablet in hand.

"We've got your results," he said, pulling up a screen full of numbers, graphs, and charts—things Amy didn't understand but suddenly wanted to. He sat across from her, glancing at the data. "Let's start with something important." He looked up. "Your body doesn't metabolize certain pain medications properly." Amy blinked. "What?"

"The pain medications you were prescribed—opioids, even some non-steroidal anti-inflammatory drugs (NSAIDs)—never had a chance to work for you. Your genetic profile shows that your liver processes them too quickly, breaking them down before they can actually provide relief." Amy sat still and thought.

All those times she had swallowed a pill, waited for relief that never came, told herself she was just being impatient. All those doctors who had suggested that she just needed to take them regularly, that the relief would build up over time. It was never going to work. For years, she had wondered if she was just being weak. If maybe the pain wasn't as bad as she thought. If she was imagining things. But it wasn't in her head—it had never been in her head. She exhaled sharply, gripping the arms of the chair. "Why didn't anyone ever test for this?"

Dr. Miller sighed. "Because most doctors don't. Standard pain treatment is trial and error; they're instructed that if one thing doesn't work, try another—without ever asking why it didn't work in the first place." Amy shook her head. She had spent years suffering, being dismissed, questioning herself—when all along, the answer had been in her biology. Pain wasn't just something she had to endure; it was something that could finally be understood.

Dr. Miller tapped on his tablet, pulling up another set of data. "There's more," he said. "Your pain isn't just from getting older, and it's not just bad luck." Amy let out a breath. She had always suspected that. But hearing someone say it—someone with proof—was different. "The real picture is more complex," he continued. "Your test results show a combination of factors that have been working against you for years." She leaned in, listening.

"First—repetitive movement injuries. Decades of the same motions, over and over. You've spent years on your feet, lifting, twisting, gripping things, handling heavy items, moving in ways that put stress on the same joints, the same muscles." Amy swallowed hard. She had always thought of her pain as something that happened to her. She had never considered that her body had been trying to keep up with the same demands, day after day, until it couldn't anymore.

"Then there's environmental exposure." He glanced at her. "You've worked in restaurants a long time, right?" Amy nodded. "Since I was a teenager." Dr. Miller continued. "Certain cleaning chemicals commonly used in commercial establishments contain compounds that can trigger chronic inflammation. Over time, exposure—even just from breathing them in—can make your body more prone to persistent pain."

Amy blinked. "You mean . . . the stuff we use to wipe down the tables and things?" He nodded. "Most people won't react. But based on your genetic markers, you do." Her chest tightened. She had spent years scrubbing, washing, breathing in those sharp, eye-watering fumes. Never once had she thought that they might be hurting her.

"And finally," Dr. Miller said, scrolling further, "your genes. You have markers that suggest a predisposition to inflammatory conditions. Some people's bodies overreact to minor injuries—what should be temporary pain lingers, becoming chronic." Amy sat back, stunned. At last, her pain wasn't a mystery. It wasn't just aging, not bad luck, not a personal failure.

She exhaled sharply, pressing a hand to her forehead. "I've spent years thinking I was just . . . weak." Dr. Miller shook his head. "You're not weak, Amy. You've just never had the right information." And for the first time in her life, she did.

A New Way Forward

Amy had expected answers to make her feel frustrated. After all these years—after all the visits that ended in shrugged shoulders and useless advice—she thought knowing the truth would just remind her of all the time she had lost. But as she sat there, listening to Dr. Miller lay everything out, she felt something completely different: relief.

Dr. Miller scrolled through his tablet. "First, we'll adjust your medication. We now know which pain medications actually work with your metabolism. No more wasting time on things your body can't use." Amy nodded, already feeling a weight lift. "Next, physical therapy," he continued. "But not generic stretches. Your pain is caused by specific movements—gripping, lifting, repetitive stress. Your therapy will focus on reversing that strain, not just stretching for the sake of stretching." That made sense. She had tried PT before, but the exercises never felt relevant to her daily life. This felt different.

"And finally," Dr. Miller said, "your work environment. Now that we know certain chemicals are triggering inflammation, we'll help you find ways to reduce your exposure. Adjusting how and when you handle cleaning products, improving ventilation—small changes that could make a big difference." Amy exhaled slowly. A clear path forward, a plan tailored to her.

Amy looked up at Dr. Miller. "So . . . what happens next?" He smiled. "Now we get to work." Hope, fragile but real, rose in her chest.

It Takes a Village

Dr. Miller pulled up the final recommendations. "Here's where we go from here," he said, his voice calm but certain. "No more guessing, no more trial and error." Amy leaned forward, listening.

"First, we're prescribing something your body can actually use. A different class of medication, one designed for people with your genetic markers. This should give you real relief—without the side effects you struggled with before." After years of pills that barely took the edge off—if they did anything at all—this was the first time someone was telling her there was a solution that would actually work for her.

"Next," he continued, "we're setting you up with targeted physical therapy. Exercises designed specifically for your repetitive motion injuries. This will focus on undoing years of strain—strengthening the muscles that need support and reducing the pressure on your joints." Amy nodded slowly. Physical therapy had always felt useless before, but this made sense.

"And finally," Dr. Miller said, "we're making adjustments to your work environment. Now that we know chemical exposure has been triggering inflammation, we'll work on reducing that. Small changes—switching to safer cleaning products, improving ventilation, even wearing protective gloves—can have a big impact on your pain levels over time.

An anti-fatigue mat, adjustable-height work surface, lighting changes—these will also help reduce the stress that's creating strain on your body." Amy felt another thing she hadn't felt in years: in control. No more doctors shrugging. No more being told to just "push through." This was the first time her pain had been treated as something real, something measurable, solvable.

"I'm submitting a request to your insurance provider," Dr. Miller added. With this data, they'll cover the treatment—no battle required." Amy blinked. No endless phone calls? No back and forth trying to prove she needed treatment? It almost felt too easy.

But when Amy picked up her new prescription and felt relief for the first time, started physical therapy, and actually noticed a difference, she knew—this was what medicine was supposed to feel like.

Conversation Redux

It happened so gradually that she almost didn't notice. One day, in the middle of a busy rush, Amy reached for the cloth to clean the tables. She had done it a thousand times before, always with a small, instinctive hesitation—bracing for pain before it came. But this time, it didn't come. She cleaned the tables easily, moving without wincing, without that sharp pull in her wrist, without the slow, creeping ache in her fingers. She paused, blinking at her hand. It wasn't just cleaning the tables. The entire day had felt . . . different.

For years, by this point in the day, she would be running on muscle memory, pushing past the pain because there was no other choice. But today, she was just working. She moved without thinking about how much it would hurt. She stepped without feeling the deep throb in her knees. She stretched her arms over her head without immediately regretting it.

And at the end of the shift? She was tired, but it was a normal kind of tired—the kind that came from a long day's work, not from fighting her own body every second. She was still standing, still moving, still herself. This time, she wasn't in pain.

It wasn't until later, while cleaning again before closing, that someone else noticed. A regular—Denise from the law office upstairs—paused before taking her bag. She paused curiously, studying Amy. "You change something?" she asked. Amy blinked. "What do you mean?" Denise shrugged, smiling. "I don't know. You just seem . . . happier."

Amy didn't know what to say. For years, she had forced every smile, every pleasant interaction, every ounce of energy that made her seem fine when she wasn't. But now she was smiling without forcing it.

Amy let out a breath, rolling her shoulders, stretching her hands—just because she could.

Before leaving, Amy brought Dr. Miller's suggestions to her manager. To her surprise, he didn't brush her off. "An anti-fatigue mat?" he repeated, looking up from the schedule. Amy nodded. "It's supposed to help with joint strain—eases the pressure on my knees and lower back." He scratched his chin, then shrugged. "Yeah, we can do that. What else?"

She hesitated. "The cleaning sprays we use . . . Dr. Miller said some of the chemicals in them could be making my and the others' pain worse. There's a different kind we can use—works just as well but without the inflammatory stuff." Her manager frowned, but not dismissively. "I'll look into it. If it's better and doesn't cost an arm and a leg"—he grinned—"absolutely." Amy hadn't expected it to be that easy.

"And one more thing," she said carefully. "Short breaks—not long, just enough to stretch a little, shift my weight. Keeps the stiffness from setting in." Her manager nodded. "You're one of the hardest workers here, Amy. If this helps keep you on your feet without being in pain, I'm all for it." She hadn't realized how much tension she had been holding in until that moment.

Amy Revitalized

Within a week, the anti-fatigue mat was behind the hostess stand. The new cleaning spray arrived in the next shipment. Amy set reminders on her phone to stretch for just two minutes every hour. They weren't huge changes, but they didn't need to be, because the difference was real. The stiffness didn't settle in as quickly. The burning in her knees eased. And at the end of the day, she didn't feel like she was dragging herself home in pieces. Small adjustments, big impact. It felt like something she could actually keep doing.

Mark noticed the change before she did.

"You're different," he said one night, stretching his legs out on the porch. Amy smiled, "Is that a good thing or a bad thing?" He laughed, shaking his head. "A good thing." Then, after a beat, "I forgot how much I missed this—you, being able to just sit and be."

Amy leaned against him, resting her head on his shoulder.

She still had moments of discomfort—some aches wouldn't disappear overnight—but she knew how to manage them. She had a plan, tools, knowledge. She was no longer at war with her own body.

One evening after work, as she was heading home, Denise from the law office passed Amy. "You really do look different these days," she said, tilting her head. "Not just happier . . . more energetic." Amy smiled, a real smile, effortless, easy. It wasn't just less pain, not just having energy at the end of the day. It was having her life back.

Amy realized that she had achieved something she never thought she'd have again: the joy of living without constant suffering. She turned back to Denise and grinned. "Yes," she said. "I am."

That night, as she and Mark strolled through their neighborhood, hand in hand, she breathed in the warm night air and let it sink in. After everything, all that she had gone through, Amy wasn't just surviving; she was living life precisely the way she wanted. Thanks to data, analysis, and precision medicine.

Pain and Personalized Therapy

Pain is one of the few experiences that unites all of humanity. Whether sharp or dull, fleeting or chronic, pain is an inevitable part of life. Yet, as Amy's story in the previous chapter illustrates, it remains one of the most misunderstood and mistreated conditions in modern medicine. Although advances in precision medicine have transformed the treatment of things like cancer and heart disease, pain management continues to lag behind, trapped in outdated models and ineffective solutions.

The scale of the problem is staggering. More than 50 million Americans suffer from chronic pain, a number that exceeds those affected by diabetes, heart disease, and cancer combined.[1] For many, pain is not just an inconvenience but a life-altering burden, limiting mobility, eroding mental health, and diminishing quality of life. It is the leading cause of disability in the United States, yet it receives far less attention and funding compared to other chronic conditions.

Despite its prevalence, chronic pain is frequently dismissed by the healthcare system. Many patients are trapped in a frustrating cycle of temporary fixes—opioids, injections, and surgeries—rather than receiving targeted, individualized care. Physicians, like Amy's primary care provider, often lack specialized training in pain management and may downplay patient complaints or prescribe treatments that fail to address the underlying causes. This failure of the system has left millions suffering in silence, seeking relief that remains just out of reach.

The Hidden Epidemic of Pain

Chronic pain is not just a medical condition; it is an economic crisis. The financial burden on the U.S. economy exceeds $560 billion annually, accounting for direct medical expenses, lost productivity, and disability-related costs.[2] These costs rival or surpass those of heart disease, diabetes, and cancer, yet pain management remains an afterthought in healthcare funding and research. Employers bear a significant share of this burden, with pain-related absenteeism and reduced work performance leading to billions in lost wages and workplace inefficiencies.

The workforce impact is equally severe. Millions of Americans are either unable to work due to pain-related disabilities or forced into early retirement. Many find themselves dependent on Social Security Disability Insurance (SSDI) or other support programs, creating a long-term economic strain on both families and government resources. Without effective treatment, pain becomes a catalyst for financial instability, pushing individuals into cycles of unemployment, reduced earning potential, and economic hardship.

Beyond financial consequences, chronic pain exacts an enormous emotional and social toll. It fractures relationships, isolates individuals from their communities, and disrupts families. The persistent struggle with pain often fuels depression, anxiety, and even substance use disorders as patients search for relief in a system that frequently fails them. The combined economic, workforce, and emotional costs of chronic pain make it one of the most overlooked yet pressing public health challenges of our time.

One-Size-Fits-All Failure in Pain Treatment

The current approach to pain management is fundamentally flawed. Instead of addressing pain proactively, the healthcare system treats it reactively, often resorting to trial-and-error methods that rely on opioids, NSAIDs, antidepressants, or anticonvulsants. These medications are prescribed with little consideration for individual differences in drug metabolism, pain sensitivity, or underlying causes. Patients are often left cycling through different medications, enduring side effects while hoping for relief that may never come.

The failure rate of traditional pain medications is alarmingly high: more than 40% of patients do not experience meaningful relief.[3] A key reason for this is biological variability. Two patients with the same diagnosis may respond completely differently to the same drug due to differences in their genetic makeup, enzyme activity, and nervous system function. Yet pain treatment still follows a generalized, one-size-fits-all model.

Consider the case of opioids. The rise of the opioid epidemic was not solely the result of patient demand—it was fueled by a healthcare system that relied on quick fixes over individualized care. In 1995, Dr. James Campbell, president of the American Pain Society, introduced the concept of pain as the "Fifth Vital Sign" (after temperature, heart rate, respiration rate, and blood pressure), encouraging doctors to assess and treat pain more aggressively.[4] Although this approach helped legitimize patient suffering, it also led to a surge in opioid prescriptions, as doctors were pressured to ensure that patients were pain-free, often without fully understanding the long-term consequences.

At the same time, pain relief became big business. Pharmaceutical companies heavily marketed opioids as safe and non-addictive despite growing evidence that long-term use led to dependency and withdrawal issues. By the early 2000s, opioid prescriptions skyrocketed, paving the way for one of the deadliest drug crises in history (see Figure 8-1).

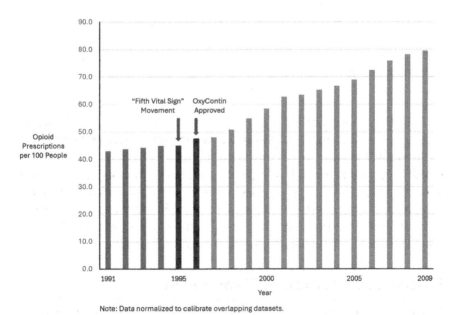

Note: Data normalized to calibrate overlapping datasets.

Figure 8-1: U.S. opioid prescriptions, 1991–2009

Source: Adapted from [5]

For years, physicians, believing them to be safe and effective, freely prescribed opioids to patients suffering from acute and chronic pain. Pharmaceutical companies assured doctors that new formulations like OxyContin were non-addictive, leading to mass opioid distribution throughout the United States. By the mid-2000s, opioid-related deaths surged, with many patients becoming physically dependent on their prescriptions. Despite mounting evidence, the damage was already done—millions were hooked, and the healthcare system had no clear solution.

In response, policy changes sought to curb the crisis by restricting opioid prescriptions. However, these sudden crackdowns had unintended consequences. Many chronic pain patients who had relied on opioids for years were abruptly cut off, with no alternative treatments in place. Facing debilitating withdrawal and untreated pain, many turned to illicit substances like heroin and fentanyl, which were cheaper and more accessible than prescription opioids. See Figure 8-2 for a timeline of the opioid crisis.

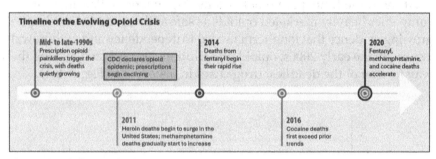

Figure 8-2: U.S. opioid crisis

Source: Adapted from [6]

This crisis could have been mitigated by precision medicine. Pharmacogenomic testing and lifestyle and environmental considerations could have identified those at high risk for ineffective treatment. Instead of blindly prescribing opioids, physicians could have pursued personalized pain management strategies, sparing thousands from addiction, withdrawal, and unnecessary suffering. The opioid epidemic underscores a painful truth: had precision medicine been widely adopted earlier, the consequences of this crisis might have been far less severe.

Misconceptions Around Chronic Pain

Chronic pain is often misunderstood—not just by society but also by the healthcare system meant to treat it. One of the greatest challenges pain sufferers face is stigma. Many patients, particularly women and minorities, report that their pain is frequently dismissed as psychological rather than physical. Instead of receiving targeted treatments, they are told to "reduce stress" or "try therapy," reinforcing a cycle of medical neglect that leaves their symptoms undiagnosed and untreated.

This bias in treatment is well-documented. Studies show that women's pain is more likely to be underestimated by doctors, leading to delayed diagnoses and inappropriate treatment.[7] Women with conditions like fibromyalgia and endometriosis often spend years searching for answers, as their symptoms are dismissed as anxiety or hormonal fluctuations. Similarly, racial disparities persist, with black and Hispanic patients less likely to receive adequate pain relief compared to white patients despite reporting the same levels of pain.[8]

What makes this even more problematic is that pain perception is not just subjective; it is biological. A person's genetic makeup influences their pain sensitivity, nervous system response, and ability to metabolize pain medications. Yet traditional medicine often fails to consider these neurological and genetic factors in treatment plans, relying instead on generalized approaches that overlook the true complexity of pain. Precision medicine offers a path forward— one that replaces bias with objective, data-driven solutions tailored to the individual. For instance, chronic pain is far more than a physical sensation: it is a multidimensional experience shaped by the nervous system, genetics, the environment, and the patient's emotional state.

> Unlike acute pain, which signals immediate injury, chronic pain rewires neural pathways, creating a cycle where the brain continues to perceive pain even in the absence of an ongoing physical cause.

Unlike acute pain, which signals immediate injury, chronic pain rewires neural pathways, creating a cycle where the brain continues to perceive pain even in the absence of an ongoing physical cause. This complexity makes treatment challenging, as traditional medicine often focuses only on symptom suppression rather than addressing the underlying neurological, genetic, and psychological factors that contribute to pain.

The connection between chronic pain and mental health is undeniable. Patients with persistent pain are at significantly higher risk for depression, anxiety, and PTSD, yet these intertwined conditions are frequently treated in isolation. A person with chronic pain may be prescribed opioids while their emotional distress is ignored, or they may be given antidepressants without any effort to identify and treat the biological sources of their pain. This fragmented approach leaves many patients cycling between specialists without meaningful relief.

Consider the veteran who suffers from both PTSD and chronic back pain. For years, doctors dismiss his symptoms as psychosomatic, assuming his pain is purely stress related. It isn't until genetic testing reveals a hyperactive inflammatory response that his doctors finally understand the physiological component of his pain. With this insight, the veteran's treatment plan can shift from ineffective medications to a targeted anti-inflammatory regimen, significantly improving his symptoms. Cases like this underscore a key lesson: Chronic pain cannot be separated from the person experiencing it. Precision medicine must address both the body and the mind.

How Precision Medicine Can Revolutionize Pain Treatment

The future of pain management lies in precision medicine: replacing guess-work with data-driven solutions. One of the most promising advancements is the use of real-world evidence to guide pharmaceutical decisions. By analyzing Big Data from millions of patients, researchers can identify patterns in how different individuals respond to specific pain medications. This means doctors can prescribe the most effective treatment from the start, moving beyond the traditional approach that leaves so many patients suffering through ineffective or harmful treatments.

Another breakthrough comes from the discovery of biomarkers for pain. New research is uncovering genetic and inflammatory markers that can predict an individual's pain sensitivity, response to opioids, and risk for chronic conditions.[9] By using genomic testing, physicians can determine whether a patient is more likely to benefit from NSAIDs, nerve blocks, or alternative therapies, ensuring that treatment is tailored to their unique biology.

But precision medicine is not just about medications. Personalized pain management also includes non-drug interventions designed to address the root causes of pain. Tailored physical therapy regimens, neuromodulation techniques (such as transcranial magnetic stimulation), and cognitive-behavioral therapy can provide long-term relief without reliance on pharmaceuticals. By integrating genetics, data analytics, and personalized care strategies, precision medicine has the potential to redefine pain treatment—offering hope where traditional medicine has failed.

NOTE   Transcranial magnetic stimulation (TMS) is a noninvasive technique that uses magnetic fields to stimulate nerve cells in the brain. It is typically used for treating depression, chronic pain, and neurological disorders.

The future of pain management must move beyond simply masking symptoms: it must focus on understanding and addressing the root causes of pain. Precision medicine offers a paradigm shift by using genetics, biomarkers, and real-world data to identify why a person experiences pain and how their body uniquely responds to treatments. Instead of relying on generalized painkillers, this approach enables targeted interventions that address the underlying biological, neurological, and inflammatory mechanisms contributing to chronic pain. To make this a reality, we must overcome ingrained barriers. Insurance companies and healthcare providers have historically been slow to adopt genetic testing and data-driven pain management strategies, often citing costs and regulatory hurdles. But the long-term benefits—reducing drug dependence, improving patient outcomes, and cutting ineffective treatments—far outweigh the upfront investment. Policymakers and payers must prioritize coverage for pharmacogenomic testing, precision diagnostics, and personalized non-drug therapies to revolutionize pain care.

For the 50 million Americans suffering from chronic pain, precision medicine offers something long overdue: hope. By replacing nonspecific, hit-or-miss treatments with tailored, effective solutions, it offers a path toward a future where pain is not a lifelong sentence but a manageable, treatable condition. The next era of pain management is one where patients are heard, treatments are personalized, and suffering is no longer the status quo.

Data as Cure

Pain treatment has relied for decades on subjective descriptions, inconsistent assessments, and a guesswork prescribing model that often fails to provide meaningful relief. Patients are asked to rate their pain on a 1-to-10 scale, yet their treatment plans are based on broad, generalized protocols rather than their individual biology. The outcome? Many suffer through ineffective medications, unnecessary procedures, and prolonged discomfort, caught in a system that treats pain as a symptom to be suppressed rather than a condition to be understood.

Modern technology is changing this, enabling a data-driven, precision-based approach. At the heart of the transformation is Big Data. By analyzing millions of patient records, genomic sequences, and real-world treatment outcomes, researchers are uncovering patterns that were previously invisible. These insights allow clinicians to predict which medications will be most effective, identify biological markers of chronic pain, and develop personalized, non-drug interventions that address pain at its source. Precision medicine is finally bringing pain treatment into the data-driven era—offering hope for those who have spent years searching for real relief.

The Data Revolution in Pain Management

Traditional pain treatment has long relied on crude, subjective measures that fail to capture the complexity of pain experiences across different patients. These assessments reduce a deeply personal, multifaceted condition to a simple number, ignoring genetic factors, nervous system function, and individual pain thresholds. This lack of precision has led to misdiagnoses, ineffective treatments, and a reliance on random prescribing, often leaving patients frustrated and underserved.

The rise of data analytics is changing this outdated model. Today, large-scale studies integrate patient-reported outcomes with biometric, genomic, and real-world evidence to uncover deeper insights into pain. Wearable devices track heart rate variability, muscle tension, and inflammatory markers, and genetic sequencing helps identify biological predispositions to chronic pain and medication responses. This shift moves pain management from symptoms to solutions. With Big Data, clinicians can now predict who is at risk for chronic pain, determine which treatments will work best for each individual, and identify patients who may be prone to medication failure or addiction. By leveraging real-world evidence and personalized analytics, pain management is engaging in an approach where treatments are targeted, effective, and tailored to each patient's unique needs.

One of the most powerful innovations in pain management is pattern recognition. By leveraging machine learning models, researchers can analyze vast amounts of patient data, from electronic health records and genomic profiles to wearable device metrics, to detect patterns in pain experiences across large populations. These insights allow clinicians to move from generalized assumptions to personalized treatment plans based on a patient's unique biological and environmental factors.

Real-world evidence is proving that pain is influenced by far more than just injury or illness. Large-scale studies have revealed that pain intensity and chronic pain risk correlate with genetics, social determinants of health, sleep patterns, and even weather conditions.[10] This level of analysis helps uncover hidden triggers that traditional medicine has long overlooked, giving doctors and patients new tools for predicting, preventing, and managing pain more effectively.

A prime example of this innovation comes from fibromyalgia research. A study utilizing Big Data analytics and biometric tracking discovered that heart rate variability, sleep disruptions, and environmental factors could accurately predict fibromyalgia flares at least one day in advance.[11] By using this information, patients were able to proactively adjust their lifestyle, increase physical therapy sessions, or modify medication regimens before symptoms worsened. This predictive capability represents a breakthrough in pain management, shifting from reactive treatments to preventative, data-driven interventions that improve quality of life.

> Big Data analytics identified that heart rate, sleep disruptions, and environmental factors accurately predict fibromyalgia events (a central nervous system processing disorder) one day in advance.

With personalized pain insights, patients can analyze how their pain fluctuates in response to diet, stress, activity levels, and environmental conditions. Advanced analytics platforms aggregate this information and compare it across defined analytic groups, allowing individuals to learn from patterns in people with similar conditions. For instance, a chronic pain patient might discover that high-sodium meals increase inflammation or that poor sleep is a predictor of next-day pain spikes—insights that were previously unavailable without data-driven tracking. With Big Data serving as the integrator and interpreter, pain management is shifting from guesswork to highly personalized, predictive science.

Similarly, integrating pharmacogenomics with Big Data offers a precision-based solution. By analyzing genetic markers linked to drug metabolism, receptor sensitivity, and pain perception, clinicians can predict which medications will be most effective and safest for each individual. This prevents unnecessary suffering, reduces adverse drug reactions, and minimizes drug dependence by ensuring that patients receive the right medication the first time.

The Gut–Brain Axis and Pain Treatment

The gut microbiome—once considered separate from the nervous system—is also recognized as a critical player in pain perception and sensitivity. Trillions of bacteria in the digestive tract influence inflammation, neurotransmitter production (such as serotonin and dopamine), and immune responses, all of which shape how individuals experience and tolerate pain. Emerging research suggests that an imbalanced microbiome can heighten pain sensitivity and contribute to chronic conditions like fibromyalgia, irritable bowel syndrome (IBS), and neuropathic pain.

This discovery has paved the way for microbiome-targeted therapies. Personalized probiotics, dietary interventions, and gut-health optimization are being explored as a means to modulate the gut–brain connection and reduce chronic pain at its source. Instead of relying solely on pharmaceuticals, precision medicine is expanding into microbiome-driven treatments that tailor gut interventions to each patient's unique bacterial composition.

A groundbreaking study recently demonstrated that specific gut bacteria influence how people metabolize common painkillers, such as NSAIDs and opioids.[12] Some individuals host bacterial strains that enhance drug efficacy, whereas others have microbes that neutralize pain medications before they can take effect. This finding opens the door to microbiome-based precision pain management, where treatments are adjusted based on a patient's gut bacterial profile.

Similarly, neurostimulation is transforming pain management by directly targeting the nervous system. Approaches such as spinal cord stimulation (SCS), vagus nerve stimulation (VNS), and TMS use electrical impulses to disrupt pain signals before they reach the brain. These implantable or noninvasive technologies modulate nerve activity, providing relief for conditions such as neuropathy, fibromyalgia, and complex regional pain syndrome (CRPS).

NOTE   Vagus nerve stimulation (VNS) is a nerve therapy that uses electrical impulses to stimulate the vagus nerve, the cranial nerve involved in regulating heart rate, digestion, and other organs.

Complex regional pain syndrome (CRPS) is a chronic pain condition that typically affects one limb (arm, leg, hand, or foot) after a stroke or heart attack. Characterized by prolonged pain disproportionate to the original injury, it often leads to significant disability.

The next frontier in neurostimulation is data-driven personalization. Traditional neurostimulators require manual adjustments, often through iterative guesswork. However, smart neurostimulators are now integrating real-time patient feedback, biometric monitoring, and data-driven analytics to fine-tune stimulation settings. By continuously analyzing pain patterns, nerve response, and patient activity levels, these devices can auto-adjust stimulation intensity and frequency—delivering optimized pain relief while preventing overstimulation and tolerance buildup.

Although neurostimulation is used to directly modulate nervous system function, integrating it with microbiome-targeted treatments such as probiotics, dietary changes, and fecal transplants can enhance pain relief, especially in conditions where inflammation is a material factor. Big Data analytics is essential in unlocking these complex relationships. By aggregating genomic data, patient-reported outcomes, and real-world evidence, researchers can identify neurologic and microbiome patterns linked to pain resilience or sensitivity. This data-driven approach ultimately allows clinicians to prescribe pain treatments based on an individual's gut composition—bringing an entirely new dimension to precision pain medicine.

Looking ahead, Big Data analytics and brainwave pattern recognition could further revolutionize pain management. Researchers are developing machine-learning models that analyze EEG and neural activity to predict pain flares before they occur.[13] Future neuromodulation systems could proactively adjust brain or spinal stimulation in real time, reducing reliance on medications and offering long-term, personalized pain relief. By combining advanced analytics with neurotechnology, precision pain medicine is moving beyond drugs toward a future where pain is managed at the source, dynamically and intelligently.

NOTE   Electroencephalography (EEG) is a technique used to record the brain's electrical activity using electrodes placed on the scalp.

A HISTORICAL LOOK AT PAIN MEDICINE'S HITS AND MISSES

Pain is as old as humanity itself. For thousands of years, people have sought ways to alleviate suffering, yet for much of history, pain management was more of an art than a science. Without an understanding of neurology, inflammation, or drug metabolism, early healers relied on superstition, folk remedies, and crude interventions—many of which did more harm than good.

Wildly Ineffective Pain Treatments of the Past
The history of pain medicine is filled with bizarre, dangerous, and wildly ineffective treatments that reflect humanity's desperation to escape suffering. Without an understanding of basic biology, healers of old often turned to theories and rituals that had no scientific basis—some of which caused more harm than the pain itself.

One of the most infamous treatments was bloodletting, mentioned in Chapter 2. Often leading to weakness or death, the practice persisted well into the 19th century despite zero medical evidence supporting its effectiveness. Equally misguided was the use of toxic substances for pain relief. Mercury, once used in ointments and pills, was believed to purge illnesses from the body but instead led to neurological damage, kidney failure, and poisoning. Arsenic and lead were also unknowingly used as pain treatments, with catastrophic consequences. Also, as Hildegard demonstrated in Chapter 6, pain was seen in many cultures as a spiritual transgression, leading to shamanistic rituals, incantations, and even exorcisms. Some believed that pain stemmed from evil spirits, and rather than seeking physical remedies, patients endured ceremonies involving chanting, fasting, or consuming hallucinogenic substances.

Despite the failures of these treatments, patients occasionally reported relief—a testament to the placebo effect. In an era with no true pain medicine, even ineffective or harmful treatments offered psychological comfort, proving that belief in a cure can sometimes be as powerful as the cure itself. These historical missteps underscored the critical need for data-driven, evidence-based pain medicine, ensuring that future treatments were guided by science, not superstition.

Breakthroughs in Pain Relief: Science Over Myth

Although pain treatment was dominated by myth and guesswork for centuries, key scientific insights eventually changed how pain was understood and managed. These discoveries replaced superstition with evidence-based medicine, ushering in a new era of effective pain relief.

One of the most significant milestones was the development of aspirin, the first modern painkiller. Derived from willow bark, which had been used in folk medicine for centuries, salicylic acid was identified as the active ingredient responsible for pain relief. In 1899, Bayer refined it into acetylsalicylic acid—aspirin—which quickly became a safe, mass-produced pain reliever (see Figure 8-3). Aspirin's ability to reduce pain, fever, and inflammation made it a cornerstone of modern medicine, proving that scientific validation could replace unreliable historical treatments.

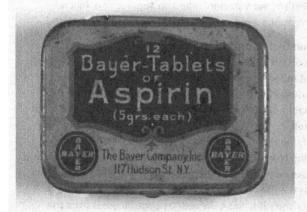

Figure 8-3: Early consumer-packaged aspirin

Source: [14] / Courtesy of Science History Institute

The 19th century also saw the rise of morphine, a game-changer in pain relief. Isolated from opium in the early 1800s, morphine was widely used during wartime surgeries to manage severe pain. However, although highly effective, its addictive potential was poorly understood, setting the stage for opioid dependence issues that persist today.

Another major breakthrough came with the discovery of anesthetics, which revolutionized surgery and childbirth. Before the mid-1800s, surgical procedures were brutal, typically performed without pain relief, and often resulted in shock or death. The introduction of ether by Boston dentist William Morton in 1846 and chloroform by Scottish physician James Simpson in 1847—who used it on himself—enabled pain-free surgeries, marking a pivotal moment in medical history.

The impact was profound. Childbirth, once a harrowing, excruciating experience, became significantly safer and more humane with the introduction of epidural anesthesia. Complex surgeries that were once unthinkable due to pain and patient trauma became routine, advancing fields such as orthopedics, neurosurgery, and transplant medicine.

NOTE   Epidural anesthesia is a type of regional anesthesia that blocks pain in a specific area of the body and is most commonly used during childbirth, surgeries, and pain management. It works by delivering anesthetic medication into the epidural space surrounding the spinal cord and nerve roots. These breakthroughs demonstrated the power of science in conquering pain, paving the way for the precision medicine revolution we see today.

The Rise of Prescription Painkillers

The 20th century followed the early insights with remarkable advancements in pain relief, introducing non-opioid and synthetic opioid painkillers that would reshape modern medicine—but also laid the groundwork for a public health crisis.

One of the most significant breakthroughs was the discovery of acetaminophen (Tylenol) and ibuprofen (Advil, Motrin). These non-opioid analgesics, developed in the mid-1900s, became widely available and highly effective for treating mild to moderate pain. They provided safe, accessible relief for headaches, arthritis, and post-surgical pain, avoiding the addictive properties of opioids. However, although effective for acute pain and inflammation, they failed to adequately address severe or chronic pain, leaving many patients without sufficient options.

To fill this gap, synthetic opioids emerged. Drugs like oxycodone (Percocet, OxyContin) and hydrocodone (Vicodin) were introduced with the promise of stronger pain relief and fewer side effects than natural opioids like morphine.

Unfortunately, these claims were misleading. Although opioids provided unparalleled relief for severe pain, their high addiction potential was downplayed or ignored—a mistake that would fuel the opioid epidemic decades later.

Lessons from History: Why Trial-and-Error Isn't Enough
Pain treatment seems to have followed a troubling historical cycle, with each generation believing it has solved pain, only to later discover the unintended consequences of its approach. Bloodletting, opium tinctures, ether, aspirin—all were once hailed as breakthroughs, yet over time, their limitations became clear. The 20th century's embrace of opioid painkillers was no different; what was hailed as a safe solution led to one of the deadliest public health crises in history.

Why does this cycle keep repeating? A key reason is that medicine has traditionally focused on treating pain after it happens rather than preventing it in the first place. Instead of identifying biological markers of pain sensitivity, medicine has relied on broad, generalized treatments that ignore individual differences. This reactive approach, often driven by pharmaceutical interests, leads to overprescription, addiction risks, and ineffective pain management.

Precision Pain Management in Action

Today, Big Data analytics is revolutionizing pain management by integrating wearable technology, genomic insights, and real-world environmental data to create a comprehensive, predictive model for pain treatment. As we saw with Amy in the previous chapter, data-driven analytics synthesizes information from multiple sources—heart rate variability, sleep quality, inflammatory biomarkers, genetic predispositions, and even environmental triggers like air quality and weather patterns—to determine how and when pain manifests in each individual.

> Medicine has traditionally focused on treating pain after it happens rather than preventing it in the first place.
>
> The solution lies in predictive, precision-based medicine. Instead of trial-and-error prescribing, doctors can now leverage Big Data, genomics, and real-world analytics to predict who will develop chronic pain, which medications will work best, and how to personalize non-drug interventions. This data-driven shift promises to break the cycle of past mistakes, ensuring that pain medicine is not just about temporary relief but about long-term, personalized solutions that prevent suffering before it begins.

This shift represents a fundamental change in how pain is treated. Instead of waiting for treatments to fail, physicians can now use data insights to anticipate patient needs, optimize medication choices, and personalize lifestyle interventions. With continuous real-world monitoring, adaptive treatment plans, and predictive modeling, pain management is moving beyond reactive prescribing to proactive, precision-driven care, ensuring that patients receive the right treatment at the right time, tailored to their unique biological and environmental realities.

One of the biggest barriers to precision pain medicine has been provider adoption, but this is beginning to change. Major health players are piloting data-driven models that determine who benefits most from precision pain analysis. By leveraging Big Data analytics, these providers can assess patient risk factors, predict long-term efforts, and justify the implementation of advanced pain management solutions.

Consider NorthShore University HealthSystem's predictive analytics in emergency services.[15] NorthShore implemented a program integrating predictive analytics directly into physicians' and nurses' workflows, combining HEART score information with patient-reported pain and health record data to determine more accurate admission needs. This initiative was developed to improve decision-making processes and patient outcomes by leveraging data-driven insights.

NOTE   The **HEART score** is a clinical tool comprising patient History, Electrocardiogram, Age, Risk factors, and Troponin levels (a cardiac biomarker) that assists emergency room (ER) personnel in stratifying the risk of individuals presenting with chest pain.[16] The HEART score is able to predict the likelihood of major adverse cardiac events over a six-week period.

A key shift is occurring in financial models as well. For instance, nearly a quarter of ER visits are for conditions that could have been treated in a lower-intensity setting. A pilot effort at Northwestern University developed a data model combining statistical methods with subjective pain information to guide mobile health interventions for chronic pain patients.[17] The model was designed to predict pain dynamics based on patient-reported pain levels and medication usage, facilitating personalized, data-driven treatment recommendations. The result? Fewer hospital admissions, more accurate medication prescriptions, and improved patient safety, all while reducing long-term healthcare expenditures.

The future of pain management is being further shaped by Big Data and precision medicine, leading to emerging therapies that customize treatment to individual biology. One of the most exciting developments is precision-designed molecules for chronic pain relief. Using data-driven computational methods like quantitative structure–activity relationship (QSAR) models, researchers are developing next-generation pain medications tailored to a patient's genetic and biochemical profile.[18] These drugs are designed to maximize effectiveness while minimizing side effects, offering an alternative to broad-spectrum pain relievers that often fail due to individual metabolic differences.

NOTE Quantitative structure–activity relationship (QSAR) is a computational modeling approach used in drug research to predict the biological activity of chemical compounds based on their molecular structure. Using QSAR modeling, scientists can learn how the chemical structure of a molecule influences its biological and pharmacological effects.

Another cutting-edge innovation is data-driven neural implants that adapt in real time to a patient's unique pain signals. Medtronic's recently approved BrainSense Adaptive stimulation device is one example, tested initially with patients with Parkinson's disease.[19] Unlike traditional spinal cord stimulators, which deliver fixed electrical impulses, these smart implants continuously adjust stimulation levels based on real-time brain and nerve activity. This allows for dynamic, highly personalized nerve signal modulation, providing relief that is far more effective than static neurostimulation methods.

These Big Data-powered innovations signal a new era in pain medicine, where treatments are dynamic, personalized, and driven by real-world patient data rather than legacy prescribing practices based on guesswork and approximation.

Cost and Outcomes of Precision Pain Management

Chronic pain doesn't just hurt the body; it destroys finances, derails careers, and fractures families. The economic toll of pain is staggering, costing the United States over half a trillion dollars annually in medical expenses, lost productivity, and disability claims.[20] For individuals, the financial burden of ineffective pain management can lead to years of out-of-pocket costs, job loss, and mental health struggles.

The current system is fundamentally broken. The guesswork approach to pain treatment forces patients to cycle through ineffective medications, unnecessary procedures, and costly specialist visits, often accumulating tens of thousands of dollars in medical debt. Meanwhile, many are prescribed opioids that fail to address the root cause of their pain, leading to dependency issues, ER visits, and further financial hardship.

Precision medicine offers a better way forward. By integrating wearables, genetic testing, and data-driven real-world analytics, physicians can identify the right treatment the first time, significantly reducing failed prescriptions, unnecessary procedures, and long-term healthcare costs. By eliminating the cycle of frustration in traditional approaches to care, precision pain medicine not only saves money but also restores lives, productivity, and hope.

The Real-World Toll on Patients and Families

Many individuals with chronic pain endure a prolonged and frustrating journey through various treatments—medications, physical therapy, and surgical procedures—often with minimal relief. This approach not only prolongs suffering but also escalates healthcare costs. Studies have highlighted the modest efficacy of many chronic pain treatments, with some clinical trials failing to demonstrate significant benefits over placebos.[21]

The economic ramifications of chronic pain are substantial. In the United States, the total incremental costs of medical expenditures for pain-related conditions are estimated to range from $261 billion to $300 billion annually.[22] Beyond direct medical expenses, patients frequently face lost productivity due to missed workdays and reduced employment opportunities. A study reported that chronic pain leads to significant reductions in family income, with households experiencing notable financial strain.[23]

As we have seen, bias in pain diagnosis and treatment is also a pervasive issue, disproportionately affecting women and racial minorities. These disparities are rooted in historical biases within medical research and practice. Traditionally, medical studies have predominantly involved white male subjects, with findings generalized to the broader population. This lack of diversity has resulted in treatment protocols that do not account for physiological and genetic differences across genders and ethnicities, leading to less effective or even harmful interventions for underrepresented groups.

Personalized pain management, through precision medicine, addresses these inequities. By leveraging technology and comprehensive patient data, healthcare providers can develop individualized treatment plans that consider a person's unique biological makeup and environmental factors. This data-driven approach can eliminate biases by focusing on objective metrics, ensuring that all patients receive effective and equitable pain management.

Finally, caring for individuals with chronic pain often imposes significant burdens on caregivers, who are frequently family members. This hidden cost of caregiving can lead to burnout, financial strain, and emotional exhaustion. A study found that caregiver burden correlates with the patient's pain intensity, depression, and lower self-efficacy.[24] This burden can manifest as physical health issues, including sleep disturbances and increased risk of chronic illnesses.

The role of caregiving often requires significant personal sacrifices as well. Many caregivers reduce their work hours or leave their jobs entirely to provide adequate support for their loved ones. This loss of income, combined with the added expenses of medical treatments and assistive devices, can lead to financial instability. The time and energy devoted to caregiving also contribute to social isolation and a decline in the caregiver's quality of life.

How Precision Medicine Identifies Simple, Effective Solutions

Precision medicine revolutionizes pain management by identifying simple, effective solutions tailored to individual needs. Wearable devices are at the forefront of this transformation, offering continuous monitoring of physiological parameters such as heart rate variability, muscle activity, and sleep patterns. These devices provide real-time data that can be analyzed to detect pain episodes as they occur, allowing for immediate and personalized interventions. For instance, electromyography (EMG) sensors in wearables can track muscle activity, providing feedback that helps in adjusting physical activity to prevent pain exacerbation.

NOTE   Electromyography (EMG) is a diagnostic procedure that measures the electrical activity of muscles in response to nerve stimulation. It helps assess nerve signal transmission, making it valuable for diagnosing neuromuscular disorders, injuries, and diseases.

The fusion of real-time data from wearables with Big Data analytics enables the development of predictive models that can identify the most effective treatments for individual patients before prescriptions are written. By analyzing patterns and correlations in large datasets, healthcare providers can match patients with therapies that are most likely to be effective based on their unique profiles.

Genetic testing has become a cornerstone, too, in preventing ineffective pain treatments. Pharmacogenomic analyses can identify genetic variations affecting drug metabolism, enabling personalized medication plans. By tailoring treatments to individual genetic profiles, hospitals can minimize ineffective prescribing, leading to cost savings and improved patient care.

Finally, advanced data analytics has unveiled significant connections between environmental factors and chronic pain. Studies indicate that exposure to air pollutants such as particulate matter correlates with increased incidences of rhinitis and sleep disturbances, which can exacerbate pain symptoms. For instance, research in Seoul, South Korea, analyzed over 93 million hospital visits and found a notable association between air pollution levels and rhinitis cases.[25]

As evidence mounts that personalized medicine can reduce healthcare costs, employers and insurance companies are beginning to adopt more proactive coverage policies. The concept of precision insurance coverage is emerging, where health plan administrators utilize data from wearable devices, genomics, and other health indicators to offer coverage tailored to an individual's specific health risks and needs.[26] This shift not only promotes preventive care but also aligns payment models with the principles of precision medicine, leading to better patient outcomes and cost savings. With these advancements, patients now have more control over their pain management than ever before. By tracking their own pain patterns, exploring personalized non-drug therapies, and advocating for precision-based treatments, individuals can take proactive steps toward more effective and tailored care. Next, we will review practical ways to incorporate precision medicine into your daily life, empowering you to move beyond outdated traditional medicine.

What You Can Do

Precision pain medicine is revolutionizing treatment, but patients play a crucial role in advocating for their own data-driven, personalized care. We've covered what's working in precision pain management broadly. Here are key steps you can take if you experience recurrent pain management challenges:

- *Empower yourself through knowledge.* Advocate for yourself and push for precision-based treatment benefits to get the most in this new era of pain medicine. Bring lifestyle and environmental data to your healthcare provider, ask questions, and request treatments tailored to your unique situation.

- *Track your pain patterns.* Using wearables or a pain journal, document how your pain fluctuates throughout the day in response to activities, diet, sleep quality, and stress levels. Wearable devices can track heart rate variability, inflammation markers, and movement patterns, providing objective data for doctors to refine treatment strategies.

- *Explore non-drug interventions.* Data-driven insights and real-world evidence suggest that diet, neurostimulation, and behavioral therapies can be highly effective. Consider the following:

  - Anti-inflammatory diets tailored to your biomarkers

  - Neurostimulation techniques, like transcutaneous electrical nerve stimulation (TENS) or vagus nerve stimulation

  - Data-guided digital therapies, such as biofeedback apps, that train your brain to modulate pain

- *Ask about pharmacogenomic testing.* Genetic testing can reveal how your body metabolizes pain medications, helping avoid ineffective drugs and dangerous side effects. Many insurers now cover pharmacogenomic testing, so ask your doctor if it's right for you.

- *Discuss new treatments.* Precision medicine is evolving rapidly. Stay updated on

  - Wearable-integrated pain management tools

  - Data-driven medication optimization

  - Personalized rehabilitation programs

By taking an active role in tracking pain patterns, exploring non-drug interventions, and leveraging precision medicine tools like pharmacogenomic testing, you can move beyond hit-or-miss treatments and toward personalized pain relief.

The integration of wearables, genetic insights, and data-driven analytics ensures that pain management is no longer limited to outdated frames of reference but is instead guided by real-world data tailored to individual needs. Those who advocate for these innovations will not only see improvements in their own pain management but also be part of a larger shift toward data-driven, proactive healthcare.

However, access to these advancements remains uneven. Although the promise of precision pain medicine is clear, not all patients have equal opportunities to benefit from these breakthroughs. Insurance reimbursement models remain outdated, often covering opioids but not genetic testing, neurostimulation, or data-driven therapies. Many physicians lack training in precision pain medicine, leading to slow adoption of these new tools. Additionally, regulatory frameworks struggle to keep pace with data-driven healthcare advancements, creating hurdles for the widespread implementation of wearable integration, genomics, and data-powered diagnostics.

In Chapter 9, we explore these critical barriers to access—from the cost of genetic testing to the slow adoption of precision medicine by healthcare providers and insurers. Understanding these obstacles is the first step toward ensuring that precision pain management is not just a privilege but the standard of care for all of us.

Notes

1. Rikard, Michaela, Strahan, Andrea, Schmit, Kristine, and Guy, Gery, "Chronic Pain Among Adults—United States, 2019–2021," *Morbidity and Mortality Weekly Report* 72, no. 15 (2023): 379–385.

   Centers for Disease Control and Prevention, "Chronic Disease Facts and Statistics," July 12 (2024), https://www.cdc.gov/chronic-disease/data-research/facts-stats/index.html.

2. Dahlhamer, James, Lucas, Jacqueline, Zelaya, Carla, Nahin, Richard, Mackey, Sean, DeBar, Lynn, Kerns, Robert, Von Korff, Michael, Porter, Linda, and Helmick, Charles, "Prevalence of Chronic Pain and High-Impact Chronic Pain Among Adults—United States, 2016," *Morbidity and Mortality Weekly Report* 67, no. 36 (2018): 1001–1006.

3. Whitten, Christine, Evans, Christine, and Cristobal, Kristene, "Pain Management Doesn't have to be a Pain: Working and Communicating Effectively with Patients who have Chronic Pain," *The Permanente Journal* 9, no. 2 (2005): 41–48.

4. Campbell, James, "APS 1995 Presidential Address," *Pain Forum* 5, no. 1 (1996): 85–88.

5. Manchikanti, Laxmaiah, Kaye, Adam, Knezevic, Nebojsa, Mcanally, Heath, Trescot, Andrea, Blank, Susan, Pampati, Vidyasagar, Abdi, Salahadin, Grider, Jay, Kaye, Alan, Manchikanti, Kavita, Cordner, Harold, Gharibo, Christopher, Harned, Michael, Albers, Sheri, Atluri, Sairam, Aydin, Steve, Bakshi, Sanjay, Barkin, Robert, Benyamin, Ramsin, Boswell, Mark, Buenaventura, Ricardo, Calodney, Aaron,

Cedeño, David, Datta, Sukdeb, Deer, Timothy, Fellows, Bert, Galan, Vincent, Grami, Vahid, Hansen, Hans, Helm, Standiford, Justiz, Rafael, Koyyalagunta, Dhanalakshmi, Malla, Yogesh, Navani, Annu, Nouri, Kent, Pasupuleti, Ramarao, Sehgal, Nalini, Silverman, Sanford, Simopoulos, Thomas, Singh, Vijay, Slavin, Konstantin, Solanki, Daneshvari, Staats, Peter, Vallejo, Ricardo, Wargo, Bradley, Watanabe, Arthur, and Hirsch, Joshua, "Responsible, Safe, and Effective Prescription of Opioids for Chronic Non-Cancer Pain: American Society of Interventional Pain Physicians (ASIPP) Guidelines," *Pain Physician* 20, no. 2 (2017): S3–S92.

Mikulic, Matej, "Rate of Prescription Opioids Dispensed in the United States from 2006 to 2022 (per 100 Persons)," May 21 (2024), https://www.statista.com/statistics/1300770/rate-of-opioid-rx-prescriptions-dispensed-in-us.

6. State Health Access Data Assistance Center (SHADAC), "The Opioid Epidemic in the United States" (2025), https://www.shadac.org/opioid-epidemic-united-states.

7. eClinicalMedicine, "Gendered Pain: A Call for Recognition and Health Equity," *EClinicalMedicine* 69 (2024): 102558.

8. Campbell, Claudia, and Edwards, Robert, "Ethnic Differences in Pain and Pain Management," *Pain Management* 2, no. 3 (2012): 219–230.

9. Kringel, Dario, and Lötsch, Jörn, "Knowledge of the Genetics of Human Pain Gained Over the Last Decade from Next-Generation Sequencing," *Pharmacological Research* 214 (2025): 107667.

10. Zorina-Lichtenwalter, Katerina, Bango, Carmen, Van Oudenhove, Lukas, Čeko, Marta, Lindquist, Martin, Grotzinger, Andrew, Keller, Matthew, Friedman, Naomi, and Wager, Tor, "Genetic Risk Shared across 24 Chronic Pain Conditions: Identification and Characterization with Genomic Structural Equation Modeling," *Pain* 164, no. 10 (2023): 2239–2252.

Beukenhorst, Anna, Schultz, David, McBeth, John, Sergeant, Jamie, and Dixon, William, "Are Weather Conditions Associated with Chronic Musculoskeletal Pain? Review of Results and Methodologies," *Pain* 161, no. 4 (2020): 668–683.

11. Dudarev, Veronica, Barral, Oswald, Radaeva, Mariia, Davis, Guy, and Enns, James, "Night Time Heart Rate Predicts Next-Day Pain in Fibromyalgia and Primary Back Pain," *Pain Reports* 9, no. 2 (2024): e1119.

12. Maseda, Damian, and Ricciotti, Emanuela, "NSAID-Gut Microbiota Interactions," *Frontiers in Pharmacology* 11 (2020): 1153.

13. Mari, Tyler, Henderson, Jessica, Maden, Michelle, Nevitt, Sarah, Duarte, Rui, and Fallon, Nicholas, "Systematic Review of the Effectiveness of Machine Learning Algorithms for Classifying Pain Intensity, Phenotype or Treatment Outcomes Using Electroencephalogram Data," *The Journal of Pain* 23, no. 3 (2022): 349–369.

14. Hoover, Gary, "The Tortuous Saga of the First Wonder Drug: Aspirin," April 3 (2020), https://americanbusinesshistory.org/the-tortuous-saga-of-the-first-wonder-drug-aspirin.

15. Olavsrud, Thor, "Healthcare Analytics: 4 Success Stories," July 13 (2020), https://www.cio.com/article/193682/healthcare-analytics-success-stories.html.

16. Six, A.J., Backus, Barbra, and Kelder, Johannes, "Chest Pain in the Emergency Room: Value of the HEART Score," *Netherlands Heart Journal* 16, no. 6 (2008): 191–196.

17. Clifton, Sara, Kang, Chaeryon, Li, Jingyi, Long, Qi, Shah, Nirmish, and Abrams, Daniel, "Hybrid Statistical and Mechanistic Mathematical Model Guides Mobile Health Intervention for Chronic Pain," *Journal of Computational Biology* 24, no. 7 (2017): 675–688.

18. Neves, Bruno, Braga, Rodolpho, Melo-Filho, Cleber, Moreira-Filho, Jose, Muratov, Eugene, and Andrade, Carolina, "QSAR-Based Virtual Screening: Advances and Applications in Drug Discovery," *Frontiers in Pharmacology* 9 (2018): 1275.

19. Park, Alice, "Parkinson's Patients Have a New Way to Manage Their Symptoms," *Time*, February 24 (2025), https://time.com/7260870/deep-brain-stimulation-parkinsons-disease.

20. Dahlhamer, James, et al., "Prevalence of Chronic Pain."

21. Colloca, Luana, "The Placebo Effect in Pain Therapies," *Annual Review of Pharmacology and Toxicology* 59 (2019): 191–211.

22. Gaskin, Darrell, and Richard, Patrick, *Relieving Pain in America: A Blueprint for Transforming Prevention, Care, Education, and Research*, National Academies Press (2011): Appendix C.

23. Kemler, Marius, and Furnée, Carina, "The Impact of Chronic Pain on Life in the Household," *Journal of Pain and Symptom Management* 23, no. 5 (2002): 433–441.

24. Tsuji, Hironori, Tetsunaga, Tomoko, Tetsunaga, Tomonori, Misawa, Haruo, Oda, Yoshiaki, Takao, Shinichiro, Nishida, Keiichiro, and Ozaki, Toshifumi, "Factors Influencing Caregiver Burden in Chronic Pain Patients: A Retrospective Study," *Medicine (Baltimore)* 101, no. 39 (2022): e30802.

25. Lee, Soyeon, Hyun, Changwan, and Lee, Minhyeok, "Machine Learning Big Data Analysis of the Impact of Air Pollutants on Rhinitis-Related Hospital Visits," *Toxics* 11, no. 8 (2023): 719.

26. Kashaboina, Murali, "The Evolving Concept of Precision Insurance Plans in Healthcare," November 8 (2024), https://www.beckershospitalreview.com/strategy/the-evolving-concept-of-precision-insurance-plans-in-healthcare.

Barriers to Access

In the preceding chapters, we have seen the liberating value of precision medicine and its more effective treatments, fewer adverse reactions, and, ultimately, better health outcomes. The technology is here, and it is already reshaping patient care.

Yet despite its promise, access to precision medicine remains uneven. The reality is that many patients cannot benefit from it due to significant barriers. Cost remains a primary obstacle, as advanced diagnostics and targeted therapies can be prohibitively expensive. Insurance coverage is inconsistent, limiting reimbursement for genetic testing and personalized treatments, and regulatory processes lag behind innovation, slowing the approval of new therapies. Meanwhile, provider adoption is slow, disparities in data inclusion persist, and public trust in genetic-driven care is still developing.

These challenges must be addressed to ensure that precision medicine is not a breakthrough for a privileged few but a revolutionary advance for all.

The Challenges of Precision Medicine

Precision medicine offers a new frontier in healthcare, but its benefits often come with an initially higher price tag than traditional methods. The tools that democratize care and make precision medicine possible create a significant cost barrier for many patients. Even wearable health devices require ongoing costs for software and data interpretation, making widespread adoption difficult for lower-income populations.

Insurance companies play a crucial role in determining who can access precision medicine.

Insurance companies play a crucial role in determining who can access precision medicine, and their hesitance to embrace cutting-edge treatments is a significant barrier. Many insurers categorize genetic testing and targeted therapies as "experimental" or "unproven," making reimbursement inconsistent. Even when coverage is available, patients often face high out-of-pocket costs due to deductibles, co-pays, and exclusions. Insurers are wary of covering expensive treatments without clear evidence of long-term cost savings, leaving many patients unable to access potentially life-saving care.

Although costs remain high today, the future holds promise for broader access. Advances in technology are driving down sequencing costs, and competition in the biotech industry is increasing affordability. Policy changes, including value-based pricing and alternative reimbursement models, can incentivize insurers to cover these treatments. As precision medicine continues to prove its long-term value, financial barriers must gradually erode, making these innovations accessible to more patients.

The Insurance Dilemma: Reimbursement Challenges and Coverage Gaps

Precision medicine, like healthcare in general, is largely dictated by the patchwork of insurance models in the United States. Coverage policies vary significantly among Medicare, Medicaid, and private insurers, leading to an inconsistent and often inequitable system.

Medicare, for instance, which serves seniors and disabled individuals, covers some precision medicine services—particularly genetic testing for certain cancers and pharmacogenomics in specific cases. However, coverage remains limited, and many emerging tests and therapies are excluded unless strong clinical utility is demonstrated. Medicaid, administered at the state level, has even greater inconsistencies; some states provide broader coverage for genetic testing, whereas others offer little to no reimbursement, making access highly dependent on geographic location.

Private insurers, although sometimes more flexible, often impose rigid preauthorization requirements and restrictions on coverage. Some insurers reimburse next-generation sequencing (NGS) for late-stage cancer patients yet deny pharmacogenomic testing for mental health or cardiovascular conditions, even when evidence confirms that it will improve treatment outcomes. As a result, the availability of precision medicine varies not based on medical need but rather on the fine print of insurance documents.

Many insurers deny claims as experimental despite FDA approvals and strong clinical backing. Even when genetic tests or targeted therapies demonstrate improved patient outcomes, insurers typically require providers and patients to navigate a complex web of appeals and authorizations. These bureaucratic roadblocks often lead to lost time: something most critically ill patients cannot afford. Faced with denials, patients are often left with two difficult choices: pay out of pocket or forgo treatment. For those without the financial means, precision medicine remains more promise than progress.

This disparity disproportionately affects low-income and underserved populations, aggravating existing healthcare inequities. Although wealthier patients may afford out-of-pocket testing or access clinical trials, others must rely on outdated, one-size-fits-all treatments that may be less effective or have severe side effects. Momentum is building to bridge the reimbursement gap, with legislators pushing for expanded Medicare and Medicaid coverage of precision therapies, recognizing their long-term cost-saving potential, and some state governments exploring mandates that require insurers to cover certain precision medicine services.

Employers are stepping in, too, with some large companies incorporating genetic testing into employee health benefits, allowing workers access to precision diagnostics and tailored treatments. Additionally, advocacy organizations such as the Personalized Medicine Coalition are actively engaging insurers, presenting data-driven arguments on the necessity of broader coverage.[1] Although progress remains slow, these steps signal a shift toward making precision medicine more accessible, and regulatory policies must evolve alongside these efforts.

Regulatory and Institutional Challenges: The Slow Pace of Adoption

The regulatory framework governing medical innovation was designed for traditional pharmaceuticals, where treatments are tested on large, diverse patient populations before receiving approval. Precision medicine, by contrast, tailors treatments to individuals based on genetic or molecular characteristics, making the standard approval process ill-suited for these innovations. The FDA's existing requirements for large-scale randomized clinical trials often create bottlenecks, delaying access to life-saving precision therapies.

Many targeted treatments, such as gene therapies and pharmacogenomic-based drugs, are only applicable to small subsets of patients, making it difficult to meet traditional statistical benchmarks for efficacy. The current system struggles to balance the need for rigorous safety assessments with the urgency of providing breakthrough treatments to those who need them most. Additionally, regulators require extensive post-market data to confirm long-term benefits, adding another layer of complexity. The challenge is finding a middle ground where treatments can be approved more rapidly without compromising patient safety.

By accelerating trials rather than changing approval thresholds, the United States has attempted to adapt its regulatory framework to accommodate this challenge. In Europe, the European Medicines Agency (EMA) has opted for lower thresholds in approving targeted therapies, often allowing conditional approvals based on promising early-phase trial data; and countries like Japan have implemented accelerated approvals for regenerative and genomic therapies, enabling faster access to innovative treatments. See Figure 9-1 for a comparison of U.S. and European accelerated approval rates.

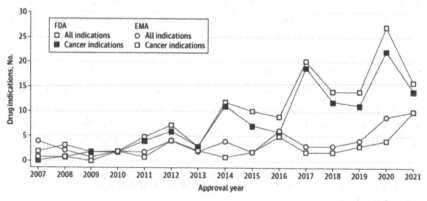

Figure 9-1: Drugs granted FDA accelerated approval and EMA conditional marketing authorization

Source: [2]/American Medical Association

Recognizing these challenges, the FDA has introduced programs to streamline approvals for precision medicine. The Breakthrough Therapy Designation and Accelerated Approval Pathway allow promising targeted therapies to move

> An obstacle to precision medicine's adoption is the lack of formal training among healthcare providers.

through the system faster. Additionally, the Real-Time Oncology Review (RTOR) process enables early data submission for certain cancer treatments, reducing the time to market.[3] Although progress is being made, continued reforms are needed to ensure that precision medicine reaches patients without unnecessary regulatory delays.

Similarly, another big obstacle to precision medicine's adoption is the lack of formal training among healthcare providers. Most practicing physicians continue to be trained in a one-size-fits-all model, with limited emphasis on genetic and molecular factors influencing disease. Although medical schools and residency programs are beginning to incorporate precision medicine into their curricula, many current providers have minimal exposure to genomic testing, pharmacogenomics, or AI-driven diagnostics.[4] As a result, they may be hesitant to integrate these advancements into patient care, leading to underutilization of precision medicine even when it is available.

Physicians already struggle with overwhelming administrative burdens, including electronic health record (EHR) management, insurance requirements, and compliance with evolving regulations. Adding genetic and molecular data to the mix can create additional complexity, especially when this information is not easily integrated into existing clinical workflows. Many providers feel ill-equipped to interpret complex genetic reports or determine how precision-based treatments fit into standard practice guidelines. Without clear, user-friendly tools, the adoption of precision medicine remains slow, as many clinicians opt for familiar, traditional treatment approaches rather than navigating a new layer of complexity.

Efforts are underway to address these challenges. Medical schools and continuing education programs are expanding coursework on genetics, AI in healthcare, and personalized treatment approaches. Institutions like the National Institutes of Health (NIH) are funding initiatives to train providers in genomic medicine.[5] Additionally, AI-driven decision support tools are emerging to help physicians integrate precision medicine into patient care. These tools analyze genetic data alongside clinical factors, offering treatment recommendations in real time. By reducing the cognitive burden on physicians, these innovations can accelerate the adoption of precision medicine and improve patient outcomes.

Data Privacy: Who Owns the Information?

The devil is in the details, as the saying goes, and in precision medicine, the details are the data. By analyzing massive datasets—genomic profiles, medical histories, wearable device metrics, and environmental factors—researchers and clinicians can develop targeted treatments tailored to individual patients. However, the ownership and control of this data remains contentious. Patients, healthcare providers, research institutions, and private companies all have a stake in this information, leading to ethical and legal dilemmas. Should genomic data belong to the patient, the lab that sequenced it, or the company that developed the analysis software? Without clear guidelines, access to critical datasets can be restricted, slowing progress in precision medicine.

Untangling the Ownership Knot—Data Regulations

Data privacy laws like the Health Insurance Portability and Accountability Act (HIPAA) are often cited as the primary gatekeepers. But HIPAA's role is narrower than many assume—and it leaves major questions unanswered, especially about who actually owns the data fueling modern care.

Enacted by the U.S. Congress in 1996, HIPAA was designed to protect the privacy and security of identifiable health information—called protected health information (PHI)—held by "covered entities" such as doctors, hospitals, insurers, and their business associates. It limits how PHI can be disclosed without patient consent, mandates data safeguards, and gives individuals the right to access and request corrections to their health records. In short, HIPAA governs access, sharing, and security. But it does not confer ownership rights to anyone—not patients, not providers, not insurers, and not labs.

HIPAA also doesn't cover all health-related data. Consumer genomics companies, fitness apps, and other direct-to-consumer platforms often fall outside its scope, even when they collect sensitive biometric data. In these cases, privacy protections are governed by company terms of service or patchwork state laws.

This contrasts with Europe's General Data Protection Regulation (GDPR), which treats personal data, including genetic data, as a fundamental right. Under GDPR, individuals can demand deletion of their data and object to certain uses, and they must give explicit consent for data sharing. Canada's Personal Information Protection and Electronic Documents Act (PIPEDA) takes a more business-friendly middle ground, requiring organizations to balance individual rights with commercial interests.

For precision medicine to mature, it's crucial to understand HIPAA's limits. Although it offers important protections, it was built for an analog age of paper records and siloed systems, not for the data-rich, cross-sector environment that precision medicine demands. Precision medicine generates data at every step, from the moment a cheek swab is collected to the final clinical interpretation. But not all data is treated the same, and neither the law nor the healthcare system draws bright lines around who controls each stage. To understand the ownership dilemma, we need to follow the data as it moves through three key phases—input, throughput, and output:

- **Inputs** are the digital data and biological materials (blood, saliva, tissue) collected from patients. Ethically, these originate from the individual and carry inherent personal identity. But once handed over for analysis, ownership becomes murky. Applications and laboratories typically require patient privacy and consent forms, transferring control over how the data and samples are used. In some cases, these inputs may be retained indefinitely for research or commercial purposes.

- **Throughputs** refer to the data derived from these samples: genetic sequences, expression profiles, or metabolomic readouts. Often referred to as *omics* data (genomics, transcriptomics, proteomics, etc.) or biomarkers, this phase includes the intermediate analytics used to generate insights but not yet shared with the clinician or patient. Because this information is often produced using proprietary software or algorithms, labs and biotech firms typically claim ownership or control over it. HIPAA provides minimal guidance here, especially if data is de-identified before use.

- **Outputs** are the clinical interpretations: a test result, a risk score, or a treatment recommendation. These typically become part of the patient's medical record and are subject to HIPAA protections, meaning patients can access them and request copies. However, patients rarely receive the full underlying data or the analysis logic that generated it, just the conclusion.

In essence, as data flows from input to output, patient rights often narrow. At each stage, different stakeholders assert control, creating a fragmented structure of data governance that often leaves patients on the outside looking in.

Aggregation and Anonymity

Individually identifiable data, like a patient's genome, test results, or clinical history, is considered PHI under HIPAA if held by a covered entity. Patients have the right to access this data and, in theory, some say over its use. But once that information is de-identified, meaning 18 specific identifiers (such as name, date of birth, and address) are removed, HIPAA no longer applies. The data becomes legally unregulated, even if it still originated from an individual.

Researchers and innovators increasingly rely on de-identified, aggregated datasets—often drawn from thousands or even millions of individuals—to power the next generation of precision medicine. These data resources are foundational to training AI models, uncovering population health trends, and accelerating the development of personalized diagnostics and therapies. Freed from personal identifiers, such datasets make it possible to recognize subtle patterns that would otherwise remain hidden, enabling medical breakthroughs at scale.

A material issue in this thorny data challenge is the representation of diverse populations in aggregated research. Historically, healthcare datasets have been disproportionately composed of individuals of European ancestry, leading to treatments that may be less effective for underrepresented racial, ethnic, and socioeconomic groups. This bias can widen health disparities, as precision medicine innovations may not be as applicable to everyone—bringing us back to the substantial value of broadly aggregated Big Data.

Not surprisingly, concerns about data security and discrimination from these challenges are growing. Patients worry about who has access to their healthcare information and how it might be used. Could insurers deny coverage based on genetic risk factors or employers use healthcare data to screen potential hires? Meanwhile, big tech companies increasingly play a role in health data collection, raising questions about the commercialization and potential misuse of personal health records. Current regulations, such as the Genetic Information Nondiscrimination Act (GINA), provide some protections, but gaps remain, particularly regarding non-insurance-based discrimination.[6]

Addressing these challenges requires stronger data governance policies. Decentralized data models, where patients retain more control over their own genetic information, can enhance transparency and trust. Efforts to improve diversity in genomic research must continue, ensuring that precision medicine benefits all populations. Finally, stricter regulations and enforcement mechanisms are needed to prevent misuse of genetic data, ensuring that technological advancements in medicine do not come at the expense of individual rights and health equity.

Building public trust in precision medicine also requires greater transparency, patient education, and stronger regulatory safeguards. Clear policies must define how genetic data is collected, stored, and shared. Informed consent processes should be robust, ensuring that patients understand how their information is used. Medical organizations and policymakers must also invest in public education initiatives to demystify genetic medicine and emphasize its benefits. By prioritizing ethical standards, equitable access, and data privacy, precision medicine can gain the trust necessary to fulfill its grand potential.

Social Determinants of Health and Access to Care

Beyond the regulatory and policy domain, social determinants of health (SDOH)—factors such as income, education, and housing—play a critical role in shaping medical outcomes. Those in underserved communities often lack access to the financial resources to afford precision-based therapies. Without addressing these factors, precision medicine risks becoming the province of a few rather than a universal solution. To fulfill its promise, SDOH considerations must be integrated into precision medicine policies, ensuring that innovation reaches all patients, not just those with the means to afford it.

The Geography of Health: Urban vs. Rural Disparities

It's a sobering fact that rural communities lack genetic specialists, precision medicine trials, and the infrastructure needed to deliver individualized care. Unlike urban centers with major academic medical institutions, rural hospitals and clinics often struggle with funding and workforce shortages, making access to genomic testing and targeted therapies extremely limited.

For rural patients, the nearest facility offering genetic testing or precision oncology services is sometimes hours away. This geographic challenge can lead to delayed diagnoses, limited treatment options, and sub-optimal health outcomes. Without local- or remote-enabled expertise, rural physicians may be less likely to recommend genomic testing, reducing patient awareness and uptake of precision medicine.

Similarly, although urban areas house top-tier medical institutions, underserved populations still face geographically related barriers. Transportation challenges, provider shortages in low-income neighborhoods, and long wait times for specialist appointments can make precision medicine just as inaccessible in parts of a city as in a rural town. Many urban hospitals serve overwhelmed patient populations, forcing individuals to navigate complex healthcare bureaucracies just to access basic services. Furthermore, lack of health literacy and distrust in the medical system—particularly among marginalized communities—can further limit engagement with precision medicine.

Fortunately, innovative solutions are emerging to address these disparities. Telemedicine allows rural patients to consult with genetic specialists remotely, reducing the need for travel, and mobile health units equipped with diagnostic tools bring genetic testing and counseling directly to underserved communities. Decentralized clinical trials, which use virtual platforms and local healthcare providers, also expand access to precision medicine research beyond major academic centers.

Education and Health Literacy

Although precision medicine relies on complex genetic data to guide treatments, the reality is that many patients struggle to interpret this information. Genetic test results often include detailed risk assessments, variant classifications, and probabilistic outcomes that require a level of health literacy beyond what most patients possess. Additionally, understanding how genomic insights translate into treatment decisions—whether through targeted therapies, pharmacogenomics, or lifestyle modifications—can be overwhelming, particularly for those with limited scientific or medical backgrounds. Many patients also lack awareness of clinical trial eligibility criteria, missing opportunities for cutting-edge treatments simply because they don't know where to look.

Health literacy disparities are compounded by inconsistent patient–provider interaction. Lower-income patients, individuals with limited education, and non-English speakers, significant delimiters of cultural health literacy, often assimilate information from healthcare providers less effectively.[7] Time constraints, implicit biases, and language barriers can result in rushed explanations or generic advice that fails to address patients' specific concerns. Figure 9-2 illustrates one example of the impact of cultural health literacy.

Many medical institutions also lack trained genetic counselors, further widening the gap between scientific advancements and patient comprehension. Without clear guidance, patients may struggle to make informed decisions about their healthcare.

To bridge the knowledge divide, culturally competent healthcare communication is essential. Medical institutions must invest in trained genetic counselors, multilingual materials, and community outreach

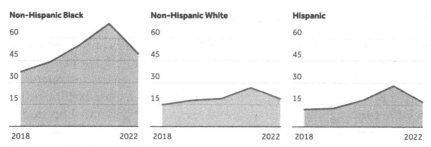

Figure 9-2: Maternal deaths per 100,000 live births by cultural cohort

Source: [8]/ PETERSON-KFF Health System Tracker

programs that explain precision medicine in accessible terms. Digital health tools, such as patient-friendly genetic report summaries and AI-powered chatbots, can help individuals understand their test results. Additionally, patient advocacy groups play a crucial role in educating communities, offering resources, and connecting individuals with healthcare providers who prioritize informed, personalized care.

To bridge the knowledge divide, culturally competent healthcare communication is essential.

At the same time, reliable internet access and digital literacy are essential for participating in technology-driven healthcare, yet millions of people—particularly those in rural areas, low-income households, and older adults—struggle with digital connectivity. Rural communities often lack broadband infrastructure, making telehealth services unreliable or entirely unavailable. Low-income families may not own smartphones or wearable devices, limiting their ability to engage with digital health programs. Older adults, who stand to benefit from remote monitoring and virtual genetic counseling, often face technological barriers, including unfamiliarity with digital platforms and difficulty navigating health apps.

Addressing this digital divide requires a multipronged approach. Public–private partnerships can expand broadband access in under-served areas, ensuring that patients can engage with virtual healthcare services; and subsidized digital health tools, such as discounted wearable devices and government-funded telehealth programs, can make precision medicine more accessible to lower-income populations. Additionally, community-based digital literacy programs, including training sessions at libraries, senior centers, and clinics, can help older adults and disad-vantaged groups navigate digital healthcare platforms.

Precision Medicine and Public Health

Although precision medicine is transforming healthcare, its adoption in public health systems remains frustratingly limited. For many uninsured and low-income patients, community health centers and safety-net hospitals serve as primary points of care. However, these facilities often lack the funding, infrastructure, and specialized personnel needed to integrate precision medicine into routine practice. Without resources for genetic counseling, next-generation sequencing, or targeted therapies, patients in these settings may only receive traditional, one-size-fits-all treatment options. This further widens the health equity gap, leaving the most vulnerable populations without access to the latest medical advancements.

Some countries are taking proactive steps to incorporate precision medicine into their public health systems. The United Kingdom's National Health Service (NHS) maintains the Genomic Medicine Service, providing support for genetic testing of conditions such as cancer and rare diseases; and Canada's national genomic initiatives aim to integrate precision diagnostics into public healthcare, ensuring broader accessibility.[9] These models demonstrate that with strategic investment and policy alignment, precision medicine can be delivered at scale.

Precision medicine has the potential to revolutionize healthcare by tailoring treatments to individual needs, but without addressing SDOH, its benefits will remain unevenly distributed. Economic status, geographic location, digital access, and systemic inequities must not be allowed to determine who can access these life-changing innovations.

A Path Forward

The promise of revolutionizing healthcare is attractive, yet high costs, insurance coverage gaps, slow regulatory approval, provider knowledge, and ethical concerns over data privacy hinder precision medicine's adoption. These challenges highlight that precision medicine is still in its early stages, requiring material changes to unlock its full potential.

To achieve this, collaboration is essential. Policymakers need to adapt regulations to accommodate personalized treatments, insurers must develop more inclusive reimbursement models, healthcare providers need better education and support, and patients must be empowered with transparency and control over their data. Only by working together can we ensure that precision medicine benefits everyone.

Despite the challenges, the future remains bright. As technology advances, costs decrease, and policies evolve, precision medicine has the potential to become an inclusive and transformative force—one that personalizes treatment, improves outcomes, and makes healthcare more equitable broadly. The takeaway is clear: precision medicine must not only advance scientifically but also advance broadly.

In Chapter 10, we bring precision medicine full circle with Susan, Michael's mother from Chapter 1, and her late-life encounter with cancer. Integrating information from the full range of innovations, we see how artificial intelligence helps Susan overcome a frightening diagnosis through the unique advantages of precision medicine.

Notes

1. Personalized Medicine Coalition, "A New Vision for Health," March 14 (2025), https://www.personalizedmedicinecoalition.org.

2. Vokinger, Kerstin, Kesselheim, Aaron, Glaus, Camille, and Hwang, Thomas, "Therapeutic Value of Drugs Granted Accelerated Approval or Conditional Marketing Authorization in the US and Europe From 2007 to 2021," *JAMA Health Forum* 3, no. 8 (2022): e222685.

3. U.S. Food and Drug Administration, "Breakthrough Therapy," January 4 (2018), https://www.fda.gov/patients/fast-track-breakthrough-therapy-accelerated-approval-priority-review/breakthrough-therapy.

   U.S. Food and Drug Administration, "Accelerated Approval," February 24 (2023), https://www.fda.gov/patients/fast-track-breakthrough-therapy-accelerated-approval-priority-review/accelerated-approval.

   U.S. Food and Drug Administration, "Real-Time Oncology Review," November 18 (2024), https://www.fda.gov/about-fda/oncology-center-excellence/real-time-oncology-review.

4. Plunkett-Rondeau, Jevon, Hyland, Katherine, and Dasgupta, Shoumita, "Training Future Physicians in the Era of Genomic Medicine: Trends in Undergraduate Medical Genetics Education," *Genetics in Medicine* 17, no.11 (2015): 927–934.

5. National Institutes of Health, "Genomic Community Resources (U24 Clinical Trial Not Allowed)," September 30 (2021), https://grants.nih.gov/grants/guide/pa-files/PAR-21-312.html.

National Institutes of Health, "NIH Awards $27M to Establish New Network of Genomics-Enabled Learning Health Systems," September 23 (2024), https://www.nih.gov/news-events/news-releases/nih-awards-27m-establish-new-network-genomics-enabled-learning-health-systems.

6. National Human Genome Research Institute, "Genetic Discrimination," January 6 (2022), https://www.genome.gov/about-genomics/policy-issues/Genetic-Discrimination.

7. Chu, Janet, Sarkar, Urmimala, Rivadeneira, Natalie, Hiatt, Robert, and Khoong, Elaine, "Impact of Language Preference and Health Literacy on Health Information-Seeking Experiences Among a Low-Income, Multilingual Cohort," *Patient Education and Counseling* 105, no. 5 (2022): 1268–1275.

8. Telesford, Imani, Cotter, Lynne, Lo, Justin, Wager, Emma, and Cox, Cynthia, "How Has the Quality of the U.S. Healthcare System Changed Over Time?" Peterson-KFF Health System Tracker, March 11 (2025), https://www.healthsystemtracker.org/chart-collection/how-has-the-quality-of-the-u-s-healthcare-system-changed-over-time.

9. Genomics England, "100,000 Genomes Project."

Innovation, Science and Economic Development Canada, "Overview of the Canadian Genomics Strategy," February 6 (2025), https://ised-isde.canada.ca/site/genomics/en.

Susan, Family, and Cancer

Susan Delaney didn't rush into things. She grew up in a town that measured time in seasons and crop reports, where school buses left dust trails on gravel roads and the best pie in town came from the church fundraiser. Her hometown in Indiana had one stoplight, one high school, and enough snow in the winter to make everyone a little tougher.

Susan was also the kind of person people naturally liked. Peggy Sue, as her friends teasingly called her, was bright-eyed and cheerful and had a gift for turning even mundane moments into fun. She remembered names, her socks were playful but paired, and her laugh made you want to laugh, too. She didn't chase the spotlight, but when something needed doing, she was already on it—usually humming a tune and making it look easy. When Susan walked into a room, it was as if the air had lightened. People felt seen. And somehow, everything just started to click.

Peggy Sue Got Married

Susan was drawn to nursing for the same reasons people trusted her: she listened carefully, moved with purpose, and didn't flinch at hard things. A practical nursing program was her next step, and it fit. She liked the early shifts, the structure, and the order of learning how to care for people in a way that made them better.

That's where she met Jack Hansen.

Jack was new to the County EMS team—tall, with big hands and an easy grin, the kind of man who fixed squeaky doors. They crossed paths during work one day. He was dropping off a patient and made a joke about the vending machine eating his last dollar. She offered him her extra granola bar, and a week later, he showed up with a replacement and asked her to lunch.

He wasn't flashy, but Jack paid attention. He remembered that she didn't like onions, that she liked cream in her coffee, and that she hummed under her breath when she was focused. Jack made her feel seen, and Susan made him feel steady. They didn't fall fast—they fit.

Jack and Susan were married the following spring in the Reformed Church on Main Street. Susan wore her cousin's wedding dress and walked down the aisle to a friend's guitar playing "In My Life." Jack's brother grilled the burgers for the reception in the church hall. There were folding chairs, casseroles, and exactly one Boerendans (farmer's dance). No one fussed, and everyone stayed late to clean up. It was perfect.

Jack was offered a job in Nebraska just after their first anniversary—a steady position with the county fire and rescue department in a growing city. Susan didn't blink. They packed their things into a borrowed trailer and drove across Illinois and Iowa, watching the landscape stretch and soften. They landed in a tidy brick house on Elm Street—a modest neighborhood shaded by old elms and buzzing with the kind of energy that made people wave at each other from driveways. Susan had the kitchen unpacked by the first weekend. She planted marigolds along the walkway and hung curtains that made the windows feel inviting. Jack built a porch swing, and Susan made the home. They mapped out their grocery store route and found a church they liked. Week by week, the house on Elm Street became theirs.

Down the street lived Amy, who worked at the café in town and always had a joke ready. Susan liked her instantly. They'd exchange zucchini in the summer, swap snow shovels in the winter, and wave from porches when neither had time to talk. Their friendship didn't require much effort—it was like a spare cup of sugar or a neighborly lookout for a package on the porch. The house on Elm Street buzzed with life. Dinner was often late, laundry never quite done, but laughter was never in short supply. Evenings brought quiet rituals—tea on the porch and Jack humming old country tunes as he turned off the lights. The days blended into each other the way good years often do—nothing flashy, but everything just right.

Susan was content. Not the passive kind of contentment, but the deep, active kind. She found joy in tending her garden, in folding warm towels, in watching her kids grow into themselves. She didn't need a different story—she was already living the one she'd hoped for.

The future felt open. There were concerts on the prairie and camping trips on the calendar. There were whispered conversations about maybe, one day, adding a sunroom or taking a road trip to the Black Hills. Nothing urgent, nothing lacking. Just more of the good.

The Family Years

Michael was born in October of Jack and Susan's third year together, red-cheeked and squinting like the sun offended him. Melissa followed two years later with bright eyes and clenched fists. Then came Danny, feet first and ready to climb. The house filled with fingerprints, mismatched socks, and cereal under every couch cushion. Susan thrived in the chaos—mothering like it was both puzzle and purpose. She didn't miss her nursing degree. Her life was full, rhythmic, and alive.

Jack worked long hours but showed up where it counted: fort-building, piggyback rides, duct-taped toys, and pancake Saturdays. The house didn't grow in square footage, but in stories—handprints on the basement wall, pencil marks on the doorframe, and a creaky board near the hallway bathroom that never did settle.

The kids left home in turn: Michael to study finance, Melissa to chase ideas and debate clubs, Danny to a music program that welcomed his scatter and his perfect pitch. Susan didn't cry, not until the doors closed. She baked too much, tidied rooms that didn't need it, and made sure their beds were always ready when they came home.

With the nest emptied, Susan and Jack found new rhythms. Long walks, shared books, and card games on chilly evenings. He built shelves; she learned to prune roses. He claimed the shed for tinkering; she took over the garage corner for her potting bench. Their life quieted but never dulled.

Michael returned to town after college and married Jennifer, a steady, organized young woman who reminded Susan of herself. Melissa moved west after a stint in Chicago. Danny bounced through music gigs before settling in Oregon. Holidays brought them all back, briefly filling the house with stories, laughter, and more voices than chairs.

When Michael and Jennifer had Noah, Susan became "Grandma Sue." She didn't hover; she helped—picking up forgotten snacks, watching Noah after school, keeping things moving. She volunteered, organized, and walked with Amy down the street. They shared recipes and town gossip, slowed their pace when Amy's knees ached, and made each other laugh over nothing at all.

Even quiet seasons have momentum. There were still to-do lists: pick up stamps, refill the birdseed, call Melissa back. Susan kept her calendar neat, her garden well-tended, and her teacups always clean and ready.

Then came that Tuesday in early spring. The wind was sharp, and the grocery list was left behind. She hadn't felt quite herself that week—tired, a little off—but she brushed it aside. She had errands to run and towels to fold.

Amy stopped by that afternoon, library book in hand together with a container of leftover soup she claimed was too much for two. They sat at the kitchen table with mugs of tea. Susan looked pale, Amy thought. "You good?" Amy asked, mid-sentence. Susan blinked, surprised. "Of course," she said, smiling. "Just a little tired." But when she stood to rinse her mug, Amy noticed the way Susan pressed her hand to her side, quickly, as if reacting to something sharp or unexpected. There was a small catch in her breath, a momentary falter. Not enough for someone else to notice, but Amy wasn't someone else.

Amy gathered the mugs, walked them to the sink, and offered to leave the soup in the fridge. But as she stepped back onto the porch, pulling her coat tighter around her, she turned once more to Susan: "You need to see the doctor." She didn't say anything else that day.

Tuesday Afternoon

The house was quiet that afternoon, the kind of quiet that waits. She leaned on the counter, steadying herself with one hand. A sudden rush of warmth flushed through her chest and up her neck. She paused—just a hot flash, maybe, or the tea with Amy catching up to her. She took a slow breath, then another, but it didn't come easily. The air felt thin, like she was breathing through fabric. She pulled out a kitchen chair and sat, resting one hand just below her ribs. There was a dull ache, not sharp, but persistent, stretching around her side and into her back. She waited for it to pass, but it didn't.

By the time Jack got home an hour later, she was folding laundry on the couch, pretending nothing was off. He noticed right away. "You okay?"

he asked, eyeing her carefully. Susan smiled, but it was the kind you put on like a jacket. "Just a little tired. Didn't sleep well." Jack walked over and placed a hand lightly on her shoulder. "You look pale." She waved him off. "I had a long day. It's nothing."

But that night, she barely touched her dinner. She winced when she reached up to the cupboard, claiming she'd "tweaked something" pulling weeds. Jack tried to believe her, tried to focus on the ballgame playing in the background, but his eyes kept drifting back to her, to the way she shifted in her seat, to the quiet way she pressed her hand to her side when she thought he wasn't looking.

The next morning, Susan noticed more. There was a strange pressure in her abdomen, subtle swelling near her collarbone, and an unfamiliar tightness beneath her ribcage that made it harder to stand for long. Still, she insisted on going through her normal routine, watering the plants, changing the sheets, and prepping soup for the freezer.

Jack stood at the sink, drying a plate that had been dry for several minutes. "Sue," he said, without turning around, "I want you to call the doctor." "I already have a checkup in a few weeks," Susan replied. "That's not soon enough." Jack persisted. She sighed. "Jack, it's probably nothing. Just age and too many muffins."

He turned then, his brow furrowed—not angry, just worried in that way men like Jack get when something doesn't add up. When something feels wrong, and you can't measure it with a wrench or a weather app. "Something's not right," he said. "And I know you. You don't sit down unless it's bad." Susan looked away, folding a dish towel and remembering Amy's comment. "I'll call tomorrow."

Jack didn't wait. He picked up the phone himself, called the clinic, and made the appointment for Friday morning. He didn't argue with her. He didn't need to. Susan saw the tremor in his voice, the way his hands wouldn't quite hold still. That night, after they turned off the lights, she reached across the mattress and took his hand. "I'll go," she said softly. "I promise."

The primary care office was familiar. Susan had been seeing Dr. Brown for years—a middle-aged, thoughtful physician who always took time to ask about Jack, about the kids, about what she was reading. This time, though, the conversation turned quickly. There were swollen lymph nodes in places they shouldn't be. Unexplained fatigue, pressure in her abdomen, subtle signs that, on their own, might mean little. But together, they drew a different kind of attention.

Dr. Brown asked more questions than usual, then paused before speaking. "I'd like to refer you to someone," she said gently. "An oncologist we work with who has advanced training in precision oncology. I'm not saying anything definitive right now. But I want someone who's trained to see the whole picture to take a look."

Susan didn't flinch. She nodded once, calm as ever, although a low hum had begun in her ears, like wind across a windowpane. Jack squeezed her hand hard. "Okay," Susan said. "We'll go."

Back at home, she brewed a cup of tea and sat in her favorite chair by the window. Jack busied himself in the kitchen, wiping counters, rearranging the spice rack, and opening and closing drawers without purpose. His restlessness had weight now. He looked strong, the same man who lifted grandkids and climbed ladders, but his eyes betrayed him. They were darker, more alert.

That night, long after Dr. Brown's words had settled like dust, Jack pulled the blanket up to Susan's shoulders and didn't say a word. He lay awake for hours, listening to the sound of her breathing, counting the pauses between each one, the way she sometimes did when their kids had the flu.

The next steps would come quickly: labs, imaging, consultations. But for now, it was still Friday. The mail sat unopened on the kitchen counter, and a half-finished puzzle lay on the dining room table. Outside, the wind stirred the chimes, and the elm tree cast a steady shadow across the porch. And in the silence between them, a new chapter waited—unwanted, unwritten, but about to begin.

A Different Kind of Doctor

The building didn't look like a cancer center. It was all glass and clean lines, more like a research lab or a startup office. There were live plants in the lobby and natural light pouring through skylights. It smelled faintly of lemon and coffee, not antiseptic. Susan found herself sitting straighter in the waiting room, like someone in a place that mattered.

Jack filled out the papers on the clipboard. Susan answered the nurse's questions in her usual clear, steady voice. She listed symptoms, dates, history, medications. But this time, the questions didn't stop there.

"Do you have any family history of cancer?"

"Have you ever had a genetic test done?"

"Would you be comfortable sharing data from your watch or fitness tracker if you use one?"

Susan blinked. "I'm sorry—my what?" The nurse smiled. "We collect a lot of data. You don't have to worry. The goal is to build a model of you, not just your symptoms."

She stepped out, and the door clicked softly shut. Jack leaned back in the chair beside her and exhaled. "Well," he said, trying for a smile, "they sure ask a lot more questions than Dr. Brown." Susan glanced at him, then returned her gaze to the blank screen on the wall. "I don't mind the questions. I just didn't know how many parts of me there were to ask about."

It wasn't fear, exactly, that hovered between them. It was a kind of reverence. The sense that something sacred was unfolding, not because of the illness, but because of the pause it created. In the silence, things had sharpened: the way Jack reached for her hand more often, the long hug from Michael that morning, the text from Melissa that read simply "Thinking of you." They were here not just because something might be wrong, but because everything was still right, and they wanted to keep it that way.

The oncologist entered with a tablet in one hand and no white coat. She introduced herself—Dr. Nia Martin—and pulled up a chair at their level. "I've reviewed your initial labs and imaging," she said. "I'll walk you through everything we see. But first, I want to know about you. Not just the chart, but you."

She asked about Susan's daily routines, her sleep, her diet, her stress, her family. She took notes not just about medications, but also about preferences. Jack watched her closely, nodding along. It felt less like an exam, more like a consultation with someone designing a map.

Dr. Martin explained that they'd send a sample for genomic sequencing. That they'd compare it against treatment pathways tailored to Susan's unique biology. That AI would assist—but not replace—judgment. "It's not just about what works," she said, "but about what works for you."

When they left, they didn't speak much. They drove home in a quiet fog, hands resting between them on the center console. At a stoplight, Jack looked over at her. "Whatever they find," he said softly, "we'll meet it head-on. Just like we always do." Susan nodded. She felt the ache in her side again, but it didn't grip her the way it had. She was already beginning to sense that the story ahead, whatever shape it took, would be one they didn't face alone.

The Machine That Thinks

Susan had always thought of machines as things with buttons, gears, or, at the very least, power cords. Something you could fix with a screwdriver or unplug if it misbehaved. But this machine—the one now thinking about her—lived in a server room hundreds of miles away, trained on the data of millions of people, drawing patterns from complexity with a precision the human brain couldn't match. And she would never see it. That, oddly, gave her comfort.

A week after her initial intake, she returned to the precision oncology center with Jack by her side. Blood tests had already been drawn, imaging completed, biopsy samples shipped overnight to a sequencing lab. What remained now was the act of waiting, although in this case, the wait was shorter than she expected.

"This isn't like the old days," Dr. Martin had said during their first meeting. "We don't have to guess anymore. We don't have to treat first and just watch."

Behind the scenes, a process had already begun, part human, part algorithm. Susan's tumor cells were being sequenced base pair by base pair, compared against a vast library of known mutations, drug responses, clinical trial data, and longitudinal studies. The AI didn't just catalog what was wrong; it predicted what would likely work, and why.

Susan didn't wear a fitness tracker or use a health app. Her life wasn't logged in steps or calories. But that didn't matter. Dr. Martin explained that environmental data, patient history, and the rhythm of her own words in the intake notes would also be parsed by natural language algorithms. Even the way she described her pain—ambiguous, dull, constant—was entered into the machine's memory, part of a growing, living model of her biology and her life.

"They're not just looking at your cancer," Jack said later, as they sat on the porch with mugs of tea. "They're looking at you. All of you." Susan nodded—that felt true.

When they returned to the center for the follow-up, the atmosphere was calm. No dramatic music, no breathless pronouncements, just Dr. Martin and her tablet, and a slim printed report clipped to a folder.

"Your results came back this morning," she said. "We've confirmed an invasive ductal carcinoma, HER2-positive. The disease has spread to at least three lymph node regions, as indicated by PET scan and biopsy correlation. Based on the current staging guidelines, this is a late Stage III, potentially Stage IV diagnosis. But . . ."

She paused, not for effect, but for clarity. "We also know exactly what we're dealing with." Jack exhaled, slow and shallow. Susan sat upright, composed, as if waiting for a recipe.

Dr. Martin explained the report: how Susan's tumor carried an amplification in the HER2 gene, a DNA mutation responsible for aggressive growth, but one now matched with effective therapies. The AI platform, trained on outcomes from over 300,000 HER2+ cases, had recommended a dual-targeted biologic treatment, combined with a de-escalated chemotherapy regimen tailored to Susan's age, cardiovascular profile, and hormone status. It had even flagged two clinical trials that Susan was eligible for: one local, another within driving distance, focused on long-term immunologic responses in HER2-positive patients post-remission.

Jack blinked. "All that from one biopsy?" "From one biopsy," Dr. Martin confirmed. "And a few terabytes of data." Susan didn't ask many questions; she listened carefully. When the word "incurable" was not spoken, and the word "treatable" was, she leaned into that space, between fear and possibility, disease and response.

"How accurate is this?" Jack asked. Dr. Martin smiled, not broadly, but with confidence. "It's not perfect. But it's better than intuition. It's built on real-world data, not just trial results. It sees combinations of biomarkers and treatment outcomes our eyes can't detect. And it learns every day—that's the difference."

Susan glanced at the folder on the desk. A dozen pages, printed in clean type, full of charts and confidence scores and small, sterile footnotes. Somewhere inside those pages was a blueprint for how her life might be saved.

A port, a "doorway" beneath the skin to deliver treatment straight into her bloodstream, would be implanted below her collarbone the following week. Treatment would begin soon after, targeted therapies first, administered every three weeks, with labs monitoring her for side effects and early response. They would reimage after four cycles and adjust as needed—nothing rigid, everything responsive.

"We've entered a new phase of cancer care," Dr. Martin said gently, closing the tablet. "It's not about just fighting. It's about understanding. We don't have to throw the whole arsenal at you, full chemo, radiation therapy. We can choose what fits."

In the parking lot, Susan stood with the folder under her arm. The spring wind was sharp, the air bright with pollen. Jack opened the car door for her but didn't speak until they were seated. "You okay?" Jack asked. She nodded slowly, "I think so."

They drove home with the radio off, the sun warming the dashboard. At a red light, Susan glanced at her reflection in the side mirror. The same face she'd worn yesterday, but something had shifted: not her fear, but her focus. The diagnosis hadn't broken her. It had clarified her. She wasn't statistics or staging. She was Susan Hansen, mother of three, wife of Jack, lover of lilies and crossword puzzles and weekly walks. She had a plan now, made by people who knew her body better than she ever could, guided by a machine that didn't just think but remembered, predicted, and learned.

Moving the Mountain

The day of Susan's first treatment, the house was quieter than usual. Jack tried to read the paper, but every headline seemed trivial. The coffee went cold before he drank it. Outside, the elm tree cast long shadows across the front walk, and the wind carried the smell of lilac from a neighbor's yard. Everything looked normal, but everything wasn't.

They arrived at the cancer center just after 9:00 a.m., checking in at a desk staffed by a nurse with bright sneakers and a voice that was both cheerful and efficient. Susan wore a cardigan and brought a book she didn't intend to open. Jack carried a thermos of tea and watched her every move.

The infusion suite wasn't what they expected. It was light-filled and oddly quiet—no harsh antiseptic smells, no urgent beeping, just the soft shuffle of carts and the occasional low murmur of nurses checking vitals. The chair reclined, and the blanket was warm. The nurse, Maria, smiled and introduced herself as if they were old friends.

"So this is the new HER2 protocol," she said, confirming the dosage and double-checking the orders. "Biologic therapy—targeted antibodies to home in on the HER2 receptors—with zero hair loss, minimal systemic side effects. It's not a walk in the park, but it's not the old sledgehammer either." Susan nodded, absorbing each word like a student before an exam. "Should I be . . . feeling something?" she asked after the first few minutes, as the medication began to enter her bloodstream. "Just rest," Maria replied. "This medicine knows what it's looking for."

At home, the family waited in a constellation of worry. Michael had taken the day off. Jennifer brought over muffins, but no one touched them. The kids, Noah, Jacob, and Emma, thought about Grandma Sue and kept asking the same question: "When can we go see her?"

Amy stopped by with a bouquet of tulips and a list of questions she didn't ask out loud. Everyone was holding their breath. But Susan didn't crash. She didn't throw up and didn't lose her appetite. In fact, she came home that night and made toast, sipped broth, and even asked Jack to put on a record while she sat on the porch wrapped in a blanket. "I'm not saying I feel great," she said, as he hovered with a worried glance, "but I feel . . . like myself."

The weeks unfolded in three-week cycles. Infusion, rest, monitor, repeat. The medicine was working, quietly, deeply, and without drama. Her scans showed progress within two months. The primary tumor was shrinking. Her lymph nodes had softened, and her bloodwork stayed steady. Side effects were there—stiff joints, some nausea, a low-grade fatigue that crept in by evening—but they were manageable, predict-able. There were no weeks lost to a low white blood cell count or risky infections, no emergency hydration drips. No frantic calls in the middle of the night.

Jack was astonished. He'd braced for battle, set his jaw for chaos, and prepared for the worst. What he got instead was a plan unfolding with calm precision. He still kept a notebook of vitals and side effects. He still checked her face in the morning before she said a word. But as the weeks turned into months, his shoulders began to relax. Slightly . . . slowly.

"She's not sick," he told Amy one morning, as they both waited for their coffee to brew in Susan's kitchen. "I mean, she is, but she's . . . not." Amy nodded. "That's what hope looks like when it's quiet."

Michael had watched the treatment plan unfold with a kind of analytical awe. He asked questions at each visit, read journal articles late at night, and built spreadsheets of trial data to understand what his mother was experiencing. But even he couldn't hide his surprise at how well she was tolerating it. "This doesn't look like what I imagined," he confessed to Jennifer one night, watching Emma help Grandma Sue with a puzzle on the living room rug. "I expected tubes, fatigue, the smell of antiseptic. Instead, she's walking in the park." Jennifer smiled, her hand on his. "She's not just being treated. She's being understood."

That, more than anything, was the shift they all felt. The care Susan received wasn't just reactive, it was responsive, built for her biology, her lifestyle, her goals. It anticipated risks and adjusted in real time. It didn't just aim to kill the cancer; it aimed to spare the rest of her.

Susan began to reclaim small pieces of her routine. She rejoined her walking group, and she cooked dinner twice a week, although she let Jack handle the clean-up. She still had days where she napped after lunch or felt the fog of fatigue settle behind her eyes. But she was living inside the treatment, not waiting for it to be over.

The grandkids learned that Grandma Sue's new medicine didn't make her lose her hair. They learned to be gentle on infusion days, and that hugs still mattered more than anything. One evening, Jacob asked if he could go to her next doctor visit. "To tell them thank you," he said, unprompted. "For fixing you." Susan knelt in front of him, her hand on his cheek. "They're helping me, sweetheart. But what's fixing me . . . is this."

The sixth cycle came and went. Then the eighth. At each follow-up, the scans improved and her tumor markers dropped. The oncologist began using the word "remission" more freely, although never casually. Susan didn't cling to the language. She clung to the feeling of climbing stairs without stopping, of laughing without effort, of planning things six months out. She knew better than to count victories too soon. But she also knew how to count the wins that mattered: a night with no pain, a full cup of coffee. A quiet moment in the garden where her daffodils, somehow, had bloomed brighter than ever.

Jack still watched her carefully, but now he smiled more and held her hand loosely instead of tightly. They resumed their old card games, started watching movies again, and talked about going back to the lake next summer. "Still no hair loss," he said one evening, brushing a stray wisp from her forehead. "Don't jinx it," she replied, laughing. "I finally like my haircut."

Amy continued to visit regularly, often arriving unannounced with fruit, stories, or just an excuse to sit. She marveled at Susan's color, her energy, and the way her voice hadn't lost its softness. She'd seen cancer before—watched it take things slowly, cruelly. But this? This was different. "They're not just fighting the cancer," she told Mark over dinner one night. "They're giving her back her life while they do it."

On a bright—different—Tuesday morning, Susan returned from her latest infusion and stepped into the kitchen to find a vase of fresh flowers on the counter. A note beside it read: "Keep blooming – Jack." She smiled, poured herself a cup of tea, and opened the window. The breeze carried in the sounds of kids playing down the block, and the faint, rhythmic thump of a basketball bouncing on the sidewalk. Her muscles ached, but she welcomed it; it meant she'd been moving. Her breath caught briefly as she reached for the tea, but she exhaled, steady. It meant she was still healing. The treatment was working. The mountain was moving.

A Life Reclaimed

Recovery, Susan discovered, wasn't something that arrived all at once. It returned in pieces, like sunlight creeping across the floor at dawn. At first, it was just energy. She woke up one day and realized she didn't need a nap by 10:00. Then her appetite returned, not for food exactly, but for flavor. One afternoon, she craved roasted garlic, the next, strawberry jam. Jack took note and stocked the fridge accordingly, smiling each time she reached for something without hesitation.

Her steps grew steadier, and her laugh returned, fuller now, from the chest. She started keeping a small list on the kitchen whiteboard, not appointments or prescriptions, but "Things That Make Me Feel Like

Me." Daffodils, homemade soup, fresh bedsheets. A walk that ended with a smile instead of soreness.

By her third clean scan, even Dr. Martin said "remission" with a little more conviction. Susan wore the word lightly, like a scarf on a breezy day, comforting, welcome, but never something she needed to prove. So when the tulips bloomed again and the air shifted into the warmth of early summer, she knew exactly what she wanted to do.

The bouquet was bright, peonies and snapdragons, wrapped in parchment paper and tied with twine. She'd picked it up that morning from the weekend market, tucking a handwritten card inside before making her way to the café. The restaurant buzzed softly with mid-morning conversation. Cutlery clinked. The espresso machine hissed. Susan stepped inside, her hair neatly brushed, her blouse crisp and cheerful. The hostess glanced up and smiled.

"Well now," Amy said, coming around from the counter, "look who's bringing the sun with her." Susan laughed and held out the bouquet. "You get the credit. You gave me the nudge." Amy took the flowers, pressing her face into them. "You didn't need a nudge. You just needed a mirror."

Behind Susan, a young woman emerged. Emily from the marketing firm upstairs glanced at Susan and Amy, caught the moment, and paused. She didn't approach, not right away, just stood there for a beat, watching Susan's easy joy, the shared laughter between the two women, the way Susan's posture filled the room without trying. It made Emily smile, one of those soft, involuntary ones that caught even her by surprise.

After a moment, she stepped forward and gave a small wave. "Hi, Mrs. Hansen." "Emily!" Susan beamed. "It's good to see you, honey." "You too," she replied. "You look—" She hesitated, then finished with a laugh. "You look amazing." "Thank you," Susan said, tilting her head warmly. "It feels good to feel good."

Emily's eyes flicked briefly to Amy, then back to Susan, a silent acknowledgment of the friendship between them, one that radiated care without asking for it. Susan didn't say anything more; she didn't need to. But she carried the moment with her as she left, the air outside full of birdsong and morning light. She walked to her car with a bounce in her step, and when she slid into the driver's seat, she hummed a little tune she couldn't quite place. Something happy, something new.

That weekend, City Park came alive. The Hansens claimed their usual spot under the big elm tree, the same one Jack had marked years ago with a small notch carved into the bark. The grandkids stretched lazily on the blanket. Susan had packed sandwiches, watermelon, lemonade, and an unreasonable number of cookies. Jennifer helped her set out the food while Michael unfolded chairs. Jack manned the cooler like a general overseeing provisions. The air was sweet with clover and distant grill smoke. An impromptu trio played music in the gazebo, and the melody drifted toward them like a breeze.

Susan watched them all with easy joy. Her eyes sparkled; her cheeks flushed from the sun. She laughed with Jacob about a soda can explosion, helped Emma tie a lopsided shoelace, and caught Jack's eye across the blanket with a look that said, "This is what we've built."

And it was. The house was still standing, the family still gathered, the sky above was still impossibly blue. No one talked about cancer. Not because they were avoiding it, but because it no longer filled the room. It had been crowded out by life. By recovery, and by the memory of the machine that had seen her fully and offered a plan precise enough to preserve her joy.

Later that afternoon, Susan lay back on the picnic blanket, arms folded behind her head, and listened to the sound of kids in the park laughing nearby. Jack sat beside her, his hand resting near hers. She didn't need to look at him to know he was smiling. The grass itched slightly, the lemonade was warm, the blanket was crooked. It was perfect.

Artificial Intelligence and the Human Genome

When Susan began her treatment journey in Chapter 10, she did so armed not only with hope but also with information—pages of lab results, genetic test data, imaging scans, and clinical notes. Yet the critical difference wasn't the amount of data she had access to, but how it was used. What made her care exceptional wasn't information alone; it was integration. An AI system absorbed her genomic sequencing, medical history, and lifestyle inputs to create a treatment plan specific to her tumor's biology and her personal health profile. For Susan, this wasn't abstract. It meant fewer side effects, greater clarity, and a path back to life. Her story introduces a pivotal idea: that in the age of precision medicine, data without synthesis is just noise.

> The challenge is no longer access to data—it is the capacity to make sense of it in a way that is timely, accurate, and personalized.

Today's healthcare system is saturated with information. Wearable devices track heart rate, blood oxygen, and sleep. Genomic sequencing identifies mutations and potential drug interactions. Electronic health records store an ever-expanding volume of structured and unstructured data. Yet without the ability to integrate and interpret these inputs, even the most advanced tools can produce confusion rather than clarity. The challenge is no longer access to data—it is the capacity to make sense of it in a way that is timely, accurate, and personalized.

This is where artificial intelligence becomes not just a tool, but an essential partner in care. AI systems can recognize patterns across genomic data, lab results, patient behaviors, and population-level outcomes, translating complexity into clinical action. In this chapter, we explore how AI transforms the genome from a string of letters into a source of insight—how it turns scattered signals into strategies, and how it augments the human capacity to heal by offering meaning where once there was only data.

The Intersection of Information and Artificial Intelligence

AI has become indispensable in making sense of the vast and complex information encoded in the human genome. Since the sequencing of the first complete human genome in 2003, the cost of sequencing has dropped from billions of dollars to less than $1,000, making widespread clinical use feasible.[1] Yet genomic data alone is not enough—it is AI that allows us to interpret this code at scale and integrate it with clinical decision-making.

The human genome contains approximately three billion base pairs and around 20,000 protein-coding genes, but identifying which variations within this code are meaningful remains a major scientific challenge.[2] Most genetic variants are harmless, whereas others are associated with disease susceptibility or drug metabolism. Some variants increase disease risk only in the presence of environmental or lifestyle factors, and others modulate treatment response. AI, particularly in the form of machine learning, enables researchers to detect clinically relevant patterns across large populations, drawing from both genetic data and associated health outcomes—an undertaking that would be virtually impossible using manual methods.

What distinguishes AI from earlier computational tools is its ability to learn and adapt from the data it processes. Rather than relying on fixed algorithms, modern machine learning systems—one way in which AI systems learn—continually refine their predictions like humans do, as new information becomes available.

> Rather than relying on fixed algorithms, modern machine learning systems continually refine their predictions like humans do, as new information becomes available.

This adaptability makes AI an ideal tool for interpreting the genome, which is not only vast in its complexity but also deeply contextual. AI is uniquely suited to integrate these dynamic variables.

Big Data plays a foundational role in powering these insights. It encompasses not only genomic sequences but also longitudinal health records, insurance claims, imaging results, laboratory values, pharmaceutical histories, and more. These diverse datasets form the training grounds for AI models that can identify subtle but statistically significant correlations between genetic variants and disease outcomes. Institutions like the UK Biobank and the Million Veteran Program have contributed to hundreds of studies leveraging this integrative approach.[3] Big Data allows AI systems to generalize across cohorts and surface predictions that would otherwise be lost in the noise of individual variation.

Big Data is also crucial in ensuring the scalability of precision medicine. Without it, AI models would lack the statistical power to recognize meaningful trends, particularly when studying rare diseases or variant–drug interactions that appear only in a small subset of the population. Furthermore, AI algorithms benefit from the inclusion of unstructured data, such as physician notes and patient-reported outcomes, which help contextualize molecular data. Natural language processing techniques now allow AI to parse these free-text fields, extracting valuable information that enhances predictive accuracy and clinical utility. This cross-pollination of data types is what makes Big Data "big" not just in volume, but also in complexity and insight potential.

Although earlier chapters explored the application of genomics in cancer and medication response, it is important to recognize the role of AI as a master integrator. Unlike traditional models that evaluate a single type of input, such as lab results or genetic tests, AI systems can synthesize data across biological, behavioral, and environmental domains. This multimodal integration is revolutionizing how we predict risk, monitor progression, and tailor interventions.

NOTE Unbound by the limits of information burden and human synthesis, AI is a master integrator across multiple domains.

One high-impact example is in oncology. HER2-positive breast cancer, characterized by amplification of the HER2 gene experienced by Susan in our earlier story, was once treated with generalized chemotherapy regimens that exposed patients to broad cytotoxic effects. Today, genomics-guided treatment regimens incorporate biologics such as trastuzumab and pertuzumab (monoclonal antibodies). Platforms like Tempus now integrate tumor sequencing, electronic health records, and population-level treatment outcomes to recommend therapy regimens with improved specificity, safety, and efficacy.[4]

What makes this especially significant is the growing ability of AI to stratify patients—not just by diagnosis, but also by underlying molecular profile and likely treatment response. This enables oncologists to tailor therapies with far greater precision, increasing the likelihood of remission while minimizing side effects. In practice, this means shorter treatment cycles, fewer adverse drug events, and improved quality of life for patients, all of which translate into lower healthcare costs and better long-term outcomes.

Pharmacogenomics is another major application of AI. A significant number of genes have documented effects on the metabolism of antidepressants, opioids, anticoagulants, and statins. AI models trained on pharmacogenomic databases have helped forecast adverse drug reactions and therapeutic failures before prescriptions are written. Incorporating this layer of genomic insight into prescribing workflows minimizes guesswork and enhances patient safety and satisfaction. These insights are only as useful as they are accessible. Increasingly, health systems are embedding AI-enhanced pharmacogenomic tools into electronic health records, enabling physicians to receive point-of-care alerts that factor in the patient's genotype and phenotypic history. A clinician treating depression, for example, may be advised in real time to avoid fluoxetine in a patient with impaired CYP2D6 function—without needing to comb through literature or consult a specialist.[5] This augmentation of clinical decision-making allows providers to focus on patient communication and care continuity, rather than acting as human data processors.

Wearable devices further expand this vision by offering continuous, real-time data on individual physiology. Heart rate variability, sleep patterns, physical activity, oxygen saturation, and even electrodermal responses can be tracked passively and transmitted to AI systems that learn personal baselines and deviations. These systems can detect flare-ups in conditions such as fibromyalgia, autoimmune disease, or congestive heart failure days before clinical symptoms prompt a doctor visit. As data from wearables becomes increasingly granular, AI serves not only as an interpreter but as a filter. The average physician cannot be expected to sift through thousands of datapoints per patient per day, but AI can. It highlights only what matters: a trend toward bradycardia, irregular circadian patterns, or reduced recovery time following exertion. AI becomes the lens that sharpens the clinician's view, surfacing actionable insights from oceans of noise.

AI's integrative role is particularly critical in complex cases. Consider a patient with multiple chronic conditions: diabetes, hypertension, and early-stage kidney disease. Each of these conditions generates streams of data—blood glucose logs, blood pressure readings, medication adherence reports, and renal function tests. AI can weave these threads into a single, coherent narrative, identifying not just what is wrong, but what matters most right now. It relieves the cognitive burden on providers while offering a more comprehensive and forward-looking view of the patient's health trajectory.

Another underappreciated strength of AI is its ability to surface non-obvious correlations. These are not associations a human clinician is trained for or expected to notice, such as connections between oral microbiome diversity and preeclampsia risk (a pregnancy complication due to high blood pressure), or the interaction between seasonal light exposure and drug metabolism. As AI models ingest more diverse datasets, including weather, geolocation, consumer behavior, and digital communication, they may generate novel insights that catalyze new research questions and health interventions. The role of the clinician, then, expands from being a diagnostician to being a curator and guide: interpreting AI-generated hypotheses through the lens of human experience and ethical care.

To be sure, integrating AI tools into existing clinical workflows is far from trivial. Most physicians are not trained in data science, and many hospitals lack the infrastructure to support real-time analytics at the point of care. Usability and clinician buy-in remain key barriers to adoption. Initiatives such as the FDA's Digital Health Software Precertification Program are working to create frameworks for evaluating the safety and efficacy of AI-based clinical decision tools.[6] Nevertheless, the long-term outlook is promising. The convergence of AI, Big Data, genomics, and wearable technology is redefining how medicine is practiced. Instead of static snapshots, like an annual checkup or a single genetic test, AI enables a fluid, responsive health profile that updates as new data becomes available. In this vision of healthcare, a patient's risk scores, treatment plans, and preventive strategies evolve over time, guided by a continuous stream of input data from their genome, environment, behavior, and body.

Looking ahead, this intersection of information and intelligence points toward the development of learning health systems—self-improving ecosystems where each patient encounter contributes to better outcomes for the next. These systems thrive on feedback loops that integrate outcomes into model refinement, creating a virtuous cycle of evidence generation and application. As Susan's story illustrated in Chapter 10, it's not just about algorithms finding the right drug—it's about preserving dignity, minimizing suffering, and reclaiming life. The genome, once viewed as a static code, becomes a dynamic component of personalized care when interpreted through AI.

Genetic and Genomic Medicine

Before delving into how genomic knowledge is transforming care, it's essential to clarify two foundational concepts that are often used interchangeably but that have distinct meanings: genetic medicine and genomic medicine. *Genetic medicine* traditionally refers to the diagnosis and management of diseases caused by mutations in a single gene. These are the so-called monogenic disorders—cystic fibrosis, sickle cell anemia, Huntington's disease—where a single genetic mutation has a well-understood, often predictable effect on disease expression.

Genomic medicine, by contrast, involves the analysis of the entire genome and the complex interplay of multiple genes, often alongside environmental factors. It aims to understand not just rare disorders but also the common, chronic conditions that affect millions—cardiovascular disease, diabetes, cancer, autoimmune and neurodegenerative disorders. Where genetic medicine identifies the needle, genomic medicine interprets the whole haystack.

This distinction is not merely academic. Genomic medicine generates vast amounts of data, creating a cognitive load that exceeds what any single physician can reasonably interpret unaided. From polygenic risk scores to pharmacogenomic variants to rare exome mutations, the clinical implications of genome-wide data are far-reaching but can easily overwhelm conventional care processes. Here is where AI plays a vital role. AI systems sift through millions of data points, highlight actionable findings, and assist clinicians in decision-making at the point of care. Without such integration, much of genomic medicine's promise risks being lost in translation.

> If the genome is the score, then AI is its conductor—interpreting, contextualizing, and orchestrating its implications in real time.

The human genome is often likened to a blueprint, but this metaphor, although helpful, is too static to fully capture the power and dynamism of our DNA. A better metaphor might be a musical score, interpreted and performed differently depending on the environment, timing, and context. If the genome is the score, then AI is its conductor—interpreting, contextualizing, and orchestrating its implications in real time. Together, the genome and AI are reshaping what medicine can do.

The Central Dogma Revisited

Since the mid-20th century, biology has operated on a central dogma: DNA makes RNA, and RNA makes protein. This flow of information—genotype to phenotype—underpins much of modern medicine. Yet recent decades have revealed that this dogma, although directionally correct, is vastly oversimplified.

Our genome contains roughly 20,000 protein-coding genes, but that's only 1–2% of the entire sequence. The remaining 98% includes regulatory regions, non-coding RNAs, and epigenetic tags: chemical modifications that affect gene expression without altering the DNA itself. These elements play critical roles in determining when, where, and how genes are turned on or off.

For example, epigenetic modifications such as DNA methylation and histone acetylation can silence or activate genes, contributing to both normal development and disease progression. Non-coding RNAs—including microRNAs and long non-coding RNAs—regulate everything from cell differentiation to immune response. And enhancer regions, located far from the genes they influence, act like molecular dimmer switches, tuning the volume of gene expression.

> **NOTE**   Methylation and acetylation are the addition of small carbon-based chemical groups to DNA that suppress or enable gene expression, respectively. By altering how tightly DNA is coiled within the cell, these modifications help control aberrant cell activity, such as cancer.

Understanding this nuanced orchestration has allowed scientists to better grasp how genetic dysfunction translates into disease. For instance, cancer is no longer seen merely as a collection of rapidly dividing cells but as a genomic disorder involving aberrant expression, regulatory failure, and epigenetic drift. The challenge is not only reading the gene code but also interpreting its language in context—a task now increasingly assisted by AI-based computational tools that analyze expression data, predict regulatory interactions, and model disease pathways.

Single-gene or monogenic disorders were the first targets of genetic medicine. Conditions like cystic fibrosis, sickle cell disease, and Huntington's disease have long been understood to arise from mutations in a single gene. What has changed is our ability to detect these mutations early, accurately, and even prenatally. Whole-exome and targeted sequencing have made it possible to identify pathogenic mutations with high sensitivity. Family screening, once based on crude pedigree charts and risk estimates, can now be conducted with genetic certainty. Newborn screening programs in the United States test for more than 30 such conditions, identifying at-risk infants before symptoms emerge—often enabling life-altering early interventions.[7]

Even more transformative are the therapeutic advances. Gene therapies, like Zolgensma for spinal muscular atrophy, offer one-time treatments that correct the underlying genetic defect. CRISPR-Cas9 has opened the door to potentially curative edits for sickle cell disease and beta-thalassemia. AI is now assisting researchers in predicting off-target effects, optimizing guide RNA design, and simulating gene-editing outcomes—making genome editing safer and more effective.

> **NOTE**   Beta-thalassemia, caused by mutations in the HBB gene, is a blood disorder that affects the body's ability to produce hemoglobin, the protein in red blood cells that carries oxygen throughout the body.

Common Diseases, Genomic Complexity

Although single-gene disorders affect millions globally, the true burden of disease lies in complex, multifactorial conditions: diabetes, heart disease, cancer, and neurodegenerative disorders. These are governed not by a single mutation but by the interplay of dozens, even hundreds of genetic variants, often influenced by environment and lifestyle.

This is where polygenic risk scores (PRS) enter the picture. PRS aggregates the effect of multiple genetic variants to estimate an individual's predisposition to a disease. They offer a genomic snapshot of risk, sometimes years before symptoms appear. For example, PRS can stratify patients for coronary artery disease with more predictive power than traditional factors like cholesterol or family history.[8]

AI algorithms are now widely used to construct and refine PRS, integrating large-scale genome-wide association study (GWAS) data with individual-level information. These machine learning models can uncover subtle genetic interactions that would otherwise go undetected by conventional statistical methods.

Yet PRS are not without limitations. Their predictive accuracy often varies by ancestry, reflecting longstanding biases in genomic databases that have historically overrepresented individuals of European descent. Environmental influences—ranging from diet and chronic stress to socioeconomic status—further complicate the picture, modulating how genetic risk manifests in ways that remain only partially understood. A high PRS for obesity, for example, may carry very different implications for someone living in a rural, low-income community versus someone in an affluent urban environment with ready access to healthcare and nutritious food.

Despite these complexities, the promise of PRS is both real and growing. In oncology, they are being used to identify women at elevated risk for breast cancer who might benefit from earlier or more frequent screening. In psychiatry, researchers are exploring how PRS could stratify patients based on their likelihood of responding to specific antidepressants—offering a more tailored path to treatment. The power of PRS lies not in deterministic prediction, but in enabling smarter, earlier intervention.

These advances are part of a broader shift—from medicine guided by general rules to one powered by dynamic, individualized learning. As traditional approaches give way to AI, the potential for precision expands exponentially.

FROM LOGIC TREES TO LEARNING MACHINES: HOW AI IN MEDICINE EVOLVED

Imagine that you're a young doctor in the 1950s. You've got a leather bag, a stethoscope, and a well-thumbed medical manual. A patient walks in with a fever, a stiff neck, and sensitivity to light. Based on your training, you recall that this triad of symptoms points to meningitis. You consult a checklist or textbook: "If fever + stiff neck + photophobia, then suspect meningitis." As depicted in Figure 11-1, that's deductive reasoning: moving from general rules to specific conclusions.

This rule-based logic, rooted in empirical observation, dominated medicine for centuries. It mirrors the structure of early AI systems, which tried to mimic human expertise using logic trees and "if-then" statements. These systems could be surprisingly effective in straightforward cases—but they had a major weakness: real life is messy.

Continues

Continued

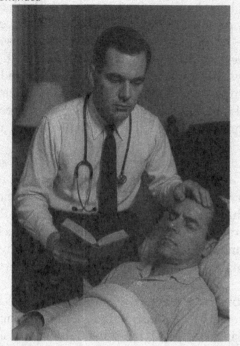

Figure 11-1: Early rules-based reasoning in medicine

Source: Generated with AI using DALL·E - OpenAI

Deductive reasoning, in many ways, built modern medicine. It gave us diagnosis manuals, treatment algorithms, and guidelines like the Framingham Risk Score, which estimates cardiovascular risk based on factors like age, cholesterol, and smoking status. It led to tools like the APGAR score for newborn health and early antibiotic guidelines for infections.

NOTE   The APGAR score is an observational test performed on a newborn baby at birth to assess their overall health and determine whether immediate medical care is needed. The name is both an acronym and a reference to its creator, Dr. Virginia Apgar, who developed the scoring system in 1952. APGAR stands for appearance (skin color), pulse (heart rate), grimace (reflex irritability), activity (muscle tone), and respiration (breathing effort), each scored on a scale from 0 to 2, with a maximum total score of 10.

Some of the first AI tools in medicine followed this same playbook. In the 1970s, a Stanford project developed MYCIN, a computer program designed to identify bacterial infections and recommend antibiotics.[9] MYCIN used a knowledge base of 450 rules. For example:

IF the site of the culture is blood,
AND the Gram stain of the organism is negative,
AND the morphology of the organism is rod,
AND the burn of the patient is serious,
THEN the organism might be Pseudomonas (illness caused by the bacteria Pseudomonas aeruginosa).

See Figure 11-2 for an example of the LISP program for this rule.

```
(defrule 52
 if (site culture is blood)
  (gram organism is neg)
  (morphology organism is rod)
  (burn patient is serious)
 then .4
  (identity organism is pseudomonas))
Rule 52:
 If
  1) THE SITE OF THE CULTURE IS BLOOD
  2) THE GRAM OF THE ORGANISM IS NEG
  3) THE MORPHOLOGY OF THE ORGANISM IS ROD
  4) THE BURN OF THE PATIENT IS SERIOUS
 Then there is weakly suggestive evidence (0.4) that
  1) THE IDENTITY OF THE ORGANISM IS PSEUDOMONAS
```

Figure 11-2: MYCIN rules-based routine
Source: [10]/with permission of Elsevier.

The system was shockingly good at identifying pathogens—even outperforming physicians in controlled settings. But it never made it into clinical use.

Why? Because MYCIN couldn't handle nuance—for example, the conclusion "might be Pseudomonas" in the rules is represented by an arbitrary 0.4 confidence value in the programming routine. It didn't adapt well to ambiguous symptoms, mixed infections, or rare cases, and worse, it couldn't explain why it reached certain conclusions beyond listing rules. Doctors, understandably, were hesitant to trust it.

Deductive systems like MYCIN assumed that the world could be cleanly divided into conditions that followed known rules. But medicine isn't always so accommodating.

A New Approach: Inductive Logic and Pattern Recognition

Today's AI takes a different path. Instead of starting with rules, it learns from data. This is inductive reasoning: generalizing from many examples to find patterns. It's how babies learn language and how you figure out your new espresso machine by pressing buttons and seeing what happens.

In medicine, this shift began with the rise of machine learning and neural networks. Rather than encoding expert knowledge manually, these systems were trained on vast datasets—millions of medical records, imaging scans, or genetic profiles. They didn't need to be told the rules. They *found* the rules.

Continues

Continued

Take cancer detection, for example. A neural network can be shown 10,000 chest X-rays labeled "cancer" or "no cancer." It doesn't know anything about tumors or lung anatomy. But after seeing enough examples, it learns to associate certain pixel patterns with cancer diagnoses.

These patterns are often too subtle for a human to see—but the machine doesn't get tired, doesn't blink, and doesn't forget. Over time, it builds a complex, weighted model. Some features matter more than others; some combinations trigger alarms. These internal weightings, represented as tensors in AI, reflect a kind of synthetic intuition.

This isn't a logic tree. It's more like a mental map, built from experience. Let's bring this into the exam room. A patient presents with chest pain. The deductive model says:

▪ Rule out heart attack → order troponins.

▪ If pain is sharp and worsens with breathing → consider pulmonary embolism.

▪ If pressure radiates to jaw and left arm → classic myocardial infarction.

A physician trained in this model runs through a mental checklist. It's fast, reliable, and familiar. But it can miss things—especially in patients with atypical presentations, like women, elderly individuals, or those with coexisting conditions.

Now imagine the inductive model:

▪ The AI reviews the patient's electronic health record (EHR).

▪ It scans their ECG, prior labs, genetics, medication history, and even real-time wearable data.

▪ It finds a 78% statistical match to a cohort of patients who had "silent" cardiac ischemia.

▪ It alerts the physician: "Consider stress testing; elevated risk despite normal troponins."

This is not deduction; it's pattern recognition on steroids. It sees correlations that no rulebook covers, and because it's updated with new data constantly, it keeps learning. See Figure 11-3 for a schematic of this pattern-recognizing machine learning model.

Gradient Descent: How Machines Learn What Matters

How do AI systems figure out what's important? One method is called gradient descent—paradoxically, a process of trial and error. Imagine a blindfolded hiker trying to find the lowest point in a valley by feeling the slope beneath her feet. She walks downhill, step by step, adjusting as she goes.

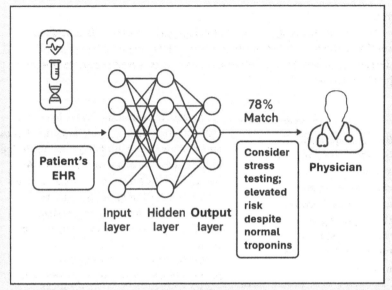

Figure 11-3: Machine learning model example

AI does something similar. It starts with a guess—a random model. It makes predictions, compares them to actual outcomes, and adjusts its internal settings (weights). After thousands or millions of iterations, it "descends" into an optimized solution.

It's slow at first. But once trained, the system can make predictions instantly. This allows AI to adapt to things humans would struggle to calculate—like how a specific gene variant modifies drug metabolism, or how 30 tiny lab deviations might signal a brewing infection.

It's not magic—it's math. And it's all insight.

Fuzzy Logic: Embracing Uncertainty

Another hallmark of modern AI is its use of fuzzy logic, where things aren't black and white. In traditional logic, something is either true or false—but fuzzy logic allows for shades of gray.

This is crucial in medicine, where symptoms are subjective and diseases often exist on a spectrum. A cough could mean nothing, or it could signal pneumonia. A genetic mutation might increase risk modestly or dramatically, depending on context.

Fuzzy logic lets AI handle that uncertainty. Instead of saying, "This is cancer" or "This is not," it says, "There's a 72% chance this is malignant." That probability can inform the next step: biopsy, monitor, or wait. It's a more realistic way to reason—because human health is rarely binary.

Continues

Continued

Despite all this progress, AI doesn't replace human clinicians. It doesn't comfort patients, ask follow-up questions, or weigh personal values in decisions. It doesn't notice when someone looks scared or says, "Something just feels wrong."

> By learning—ironically—from trial and error, across a very large number of circumstances, and applying the knowledge at the beginning of a treatment plan, AI minimizes hit-or-miss medicine.

Instead, by learning—ironically—from trial and error, across a very large number of circumstances, and applying the knowledge at the beginning of a treatment plan, AI minimizes hit-or-miss medicine and makes doctors better. It augments memory, narrows diagnoses, flags unseen risks, and sifts through more data than any human could.

The best care comes from this partnership: human empathy + machine intelligence. That's not science fiction, and it's happening now.

A Shift in How We Think

AI in medicine isn't just a new tool. It represents a fundamental shift—from rule-following to pattern-discovering, from deductive logic to inductive learning. It mirrors how great clinicians think: not rigidly, but intuitively, shaped by experience. Now machines are beginning to do the same.

In this way, AI isn't replacing the art of medicine. It's teaching us to speak its language.

Traditionally, diagnoses have been based on symptoms, imaging, and pathology. But genomic medicine is redefining diagnosis as a molecular fingerprint. Instead of labeling a patient with "breast cancer," we now distinguish between subtypes like HER2-positive, estrogen receptor-negative, or BRCA1-mutated triple-negative breast cancer.

This molecular stratification allows for targeted therapies. Patients with EGFR-mutated lung cancer respond to tyrosine kinase inhibitors; those with BRAF-mutant melanoma benefit from mitogen-activated protein kinase (MAPK) pathway inhibitors. In rare and undiagnosed diseases, whole-genome sequencing can reveal the culprit gene—sometimes solving diagnostic odysseys that have lasted years.

NOTE Tyrosine kinase inhibitors (TKIs) and mitogen-activated protein kinase (MAPK) inhibitors are types of targeted cancer therapies that block the action of enzymes called kinases, which play a crucial role in the signaling pathways that regulate cell division, growth, and survival.

BRAF is a gene that makes a protein called B-Raf, part of the MAPK signaling pathway, which, when the gene is mutated, can drive uncontrolled cancer cell proliferation.

AI-driven diagnostic platforms are increasingly being used to match genomic data with phenotypic profiles, accelerating the time to diagnosis and helping clinicians interpret ambiguous or rare variants. These systems can flag patterns across millions of data points, offering insights that might escape human analysis.

Even in infectious disease, genomics enables precision. Sequencing the pathogen's genome helps determine resistance patterns and optimal treatment. The COVID-19 pandemic underscored this utility, with genomic surveillance guiding vaccine updates and public health strategy.

Gene Therapy and Editing: Engineering the Genome

Beyond diagnosis and targeted therapy lies a more audacious goal: editing the genome itself. Gene therapy, delivering corrected genes into patient cells, has progressed from failed 1990s trials to today's FDA-approved treatments. In spinal muscular atrophy, a single infusion of Zolgensma delivers a functioning SMN1 gene, halting disease progression in infants.

> **NOTE**   The SMN1 gene (Survival Motor Neuron 1) produces a protein essential for the maintenance of motor neurons, and mutations or deletions in this gene cause spinal muscular atrophy, a genetic disorder that leads to progressive muscle weakness and atrophy.

CRISPR-Cas9 technology, first described in 2012, allows precise edits at the DNA level. In 2023, the UK approved the world's first CRISPR-based treatment for sickle cell disease, marking a historic moment in medicine.[11] Other techniques like base editing and prime editing aim to improve specificity and reduce off-target effects.

AI is instrumental in guiding these developments—from optimizing Cas9 enzyme variants and minimizing off-target activity to selecting ideal delivery mechanisms and patient cohorts for clinical trials. Predictive models help refine which edits will be most effective, and synthetic biology platforms simulate outcomes before they are tested in living systems.

Despite remarkable progress, challenges remain. Delivering genes safely and effectively, especially to hard-to-reach tissues like the brain, is difficult. Mosaicism, where only some cells are edited, can limit efficacy. Ethical concerns also loom large: germline editing, which affects future generations, raises questions about consent, equity, and eugenics.

Yet the therapeutic potential is staggering. Inherited retinal disorders, hemophilia, muscular dystrophy—conditions once deemed untreatable—are now targets of clinical trials. As costs decrease and safety improves, gene editing may soon expand from rare diseases to common ones.

The arc of genomic medicine bends from understanding to action. Sequencing the genome is not an end in itself. The goal is intervention—timely, targeted, and tailored. It's not enough to know a patient carries a BRCA mutation; we must act on it through early screening, prophylactic surgery, or novel therapies.

AI is accelerating this translation, helping clinicians prioritize variants, identify actionable findings, and match patients to trials or treatments. These algorithms can mine the literature, flag drug–gene interactions, and even predict future disease trajectories based on evolving patient data. The value of genomic insight lies in its clinical consequence: that is, who gets treated, when, and with what—and therein lies the value of personalized treatments.

Cost and Outcomes: Advanced Personalized Treatments

Powered by AI, precision medicine offers not just scientific innovation but also the promise of delivering more care with fewer resources. Traditional healthcare systems rely on generalized protocols, built on population averages, to treat diverse individuals. This one-size-fits-all approach yields acceptable results for many, but it also generates waste, adverse effects, and missed opportunities for early intervention.

AI-enhanced precision medicine flips this script. By tailoring prevention and treatment to a patient's unique biology—integrating genomics, biomarkers, behavior, and environment—care becomes proactive rather than reactive. In the following sections, we explore several categories of care where this approach is already altering costs and outcomes, supported by demonstrated examples of implementation and impact.

Acute Care

Acute care settings, where timing matters and mistakes accelerate cost, are natural proving grounds for AI-driven precision medicine. Here, personalized approaches can yield immediate improvements: better risk prediction, more targeted interventions, faster recoveries, and fewer unnecessary procedures. Across the domains of cancer treatment, cardiovascular disease, pharmacogenomics, and hospital operations, precision technologies are reshaping how clinicians triage, diagnose, and treat patients under critical conditions. We begin with perhaps the most visible and transformative example: precision oncology.

Cancer Care: Targeted Therapy

As we saw with Susan's story in Chapter 10, few areas of medicine showcase the value of precision more dramatically than oncology. Traditionally, cancer treatment relied on broad anatomical categories (e.g., breast cancer, lung cancer) and used chemotherapy regimens with punishing side effects. Success depended more on averages than individual response.

Contrast this with today's targeted therapies. At institutions like Memorial Sloan Kettering Cancer Center (MSK), tumor sequencing is standard for most advanced cancer patients. AI systems such as MSK's Watson for Oncology have been used to match patients with molecularly targeted therapies based on tumor genomics.[12]

Take non-small cell lung cancer (NSCLC): patients with EGFR mutations respond far better to TKIs than to chemotherapy. A 2020 cost-effectiveness study found that targeted therapy for EGFR-positive NSCLC patients improved median survival over traditional treatment by four to six months while reducing severe adverse events requiring hospitalization by 30%, offsetting the high drug cost.[13]

Similarly, breast cancer patients with BRCA1/2 mutations can benefit from PARP inhibitors, which reduce recurrence risk and are associated with higher quality-adjusted life years (QALYs) compared to traditional regimens.

NOTE A PARP inhibitor is a type of drug that blocks the activity of poly (ADP-ribose) polymerase (PARP) enzymes, which help repair damaged DNA in cells. Cancer cells, especially those with BRCA1 or BRCA2 mutations, have broken DNA repair pathways, and PARP-inhibiting drugs prevent cancer cells from repairing and replicating.

AI enhances this approach by rapidly analyzing sequencing data, comparing molecular profiles to trial databases, and offering evidence-based treatment options faster than traditional tumor boards.

Cardiovascular Disease: Predictive Prevention

Cardiovascular care has historically relied on generalized tools like the Framingham Risk Score to guide prevention. These models often underperform in diverse populations and fail to identify high-risk individuals early enough.

Enter AI-assisted genomic risk scoring. PRS combine hundreds of genetic markers to estimate an individual's lifelong risk of conditions like coronary artery disease (CAD). In a 2018 study published in Nature Genetics, PRS identified patients with threefold higher CAD risk even when traditional measures (like LDL levels) appeared normal.[14]

Some health systems, like Intermountain Healthcare, are beginning to combine PRS with imaging tools such as coronary artery calcium (CAC) scoring and wearable data to guide early intervention. When integrated with AI models trained on EHRs, these tools allow clinicians to tailor statin therapy more effectively.[15]

Cost comparisons show promise: research programs using AI and leading risk score indicators enhanced net reclassification improvement (NRI) by 5–10% and improved misdiagnosis by 47%, reducing associated adverse event, physician visit, and hospital admission costs by an estimated $644,000 per 100,000 patients over traditional methods.[16]

Pharmacogenomics: Fewer Failures

As we saw with Emily's story in Chapter 4, traditional prescribing often follows a trial-and-error method of implementation. In fields like psychiatry or cardiology, patients may cycle through multiple medications before finding one that works. This process is inefficient and costly, emotionally and financially.

Pharmacogenomics changes that. Using genetic tests to identify how patients metabolize drugs, clinicians can avoid ineffective or harmful choices. AI takes this further by interpreting complex drug–gene interactions and flagging contraindications within EHR systems.

Vanderbilt University Medical Center was an early pioneer, creating the Pharmacogenomic Resource for Enhanced Decisions in Care & Treatment (PREDICT) program.[17] Early results confirmed that more than 91% of tested patients revealed at least one high-risk genotype for medication failure, and the percentage continues to rise with increasing pharmacogenomic research. Moreover, the efficacy of AI-assisted predictive use of pharmacogenomic integration has been confirmed to demonstrate cost savings on the order of $1,000 to $4,000 per patient per year.[18]

Similarly, the Mayo Clinic has implemented AI-informed pharmacoge-nomics across the spectrum of patient care with its Center for Individual-ized Medicine.[19] Closely related research has shown a 43% lower hospital readmission rate experienced when using predictive pharmacogenomic systems, and a corresponding reduction in medication-related compli-cations.[20] See Figure 11-4 for an overview of the improved outcomes resulting from various integration efforts.

Author (Year), Country	All-cause hospitalization (%)	Emergency department visit (%)	Outpatient visit (%)
Bnxner et al. (2016), US	Risk was reduced by 39% relative to the control group. (RR = 0.61)	Risk was reduced by 71% relative to the control group. (RR = 0.29)	Outpatient visits increased by 97% relative to the control group. (RR = 1.97)
Eliott et al. (2017), US	Risk was reduced by 52% relative to control group. (RR = 0.48)	Risk was reduced by 42% relative to control group. (RR = 0.58)	Not measured
Epstein et al. (2010), US	31% less likely to be hospitalized than patients in the control group. (HR = 0.69)	Not measured	Not measured
Perls et al. (2018), US	58% difference between the intervention and control group. (RR = 0.42)	40% difference between the intervention and control group.	13% difference between the intervention and control group.
Ruaño et al. (2020), US	No significant difference between the intervention and control group.	Not measured	Not measured

Figure 11-4: Overview of clinical integration of predictive pharmacogenomic testing[21]
Source: [21]/Frontiers Media/CC BY 4.0

These outcomes translate into better patient adherence, fewer ER visits, and overall cost savings in high-utilization populations.

Hospital Operations: Predictive AI in Inpatient Settings

Beyond diagnosis and treatment, AI can drive efficiencies in hospital logistics. Predictive analytics models are now used to anticipate bed requirements, identify patients at risk for certain conditions, and reduce unnecessary admissions.

At Mount Sinai Health System, an early AI tool called DeepPatient combined EHR data with other clinical presentations in a machine learning network that was able to predict diagnoses of 78 different conditions with 77% accuracy.[22] Although far from perfect, this early integration of data with AI processes has allowed Mount Sinai to improve and expand the application of artificial intelligence across its hospital operations, culmi-nating in the establishment of the Center for AI and Human Health to accelerate and coordinate AI use.[23] Similarly, Cleveland Clinic's use of AI-powered bed management tools, supported by Palantir Technologies, achieved a 75% reduction in time spent calculating bed capacity, more than a 10% increase in daily hospital transfer admissions, a 20% increase in bed assignment efficiency, and approximately a one-hour reduction in daily emergency department hold time per patient.[24]

When paired with genomic risk insights, these predictive systems can also help prioritize who needs urgent genomic testing or tailored discharge planning.

Complex Conditions and Continuum of Care

Acute care innovations aim to solve immediate threats to life and health, but many of the most costly and challenging burdens in healthcare unfold over time. Complex conditions like rare diseases and behavioral health disorders do not follow predictable trajectories. They demand sustained precision, navigating uncertainty, personalizing support, and adapting care across months or even years. The following examples illustrate how precision medicine is beginning to transform the continuum of care for these patients, offering faster answers, targeted interventions, and hope where traditional models fall short.

Rare and Undiagnosed Diseases

Families seeking answers for rare diseases often endure years of misdiagnoses and unnecessary tests. The financial and emotional costs are staggering. A 2023 study found that the average diagnostic odyssey lasts more than six years and costs an average of more than $200,000 per patient.[25]

To address this challenge, AI platforms like Face2Gene have been developed to match facial phenotypes with known syndromes, accelerating recognition of rare disorders. Achieving success rates of 56% to 100%, depending on the syndrome in a universe of 360 syndromes, Face2Gene shortens the diagnostic odyssey and generates proportionate cost savings. Face2Gene is available for free to healthcare providers.[26]

In addition, the Undiagnosed Diseases Network (UDN), supported by the NIH, uses whole-genome sequencing combined with AI-enhanced variant interpretation. Since its inception in 2015, the UDN has diagnosed 855 previously unsolved cases through collaborative analysis by 24 participating health systems, and 503 novel cases have been evaluated. Novel cases are referred to the network's Model Organism Screening Center, using genomic replication in fruit flies and other organisms to identify and model clinical treatments.[27]

Although sequencing costs have dropped dramatically, the added value comes from AI tools that reduce the interpretive burden and speed up clinical decision-making.

Behavioral Health

Mental health services are overwhelmed by demand, often delayed by months-long waitlists and limited to medication-driven approaches—a fertile ground for AI-driven improvements.

AI-based solutions such as Woebot and Wysa use natural language processing (NLP) to deliver conversational cognitive behavioral therapy (CBT) via chat interfaces. Although not a replacement for licensed therapists, they serve as early triage and support tools.

> **NOTE**   Natural language processing (NLP) is a branch of AI that enables computers to understand, interpret, and generate human language. It combines linguistics, computer science, and machine learning to process and analyze natural language data.
>
> Cognitive behavioral therapy (CBT) is a structured, goal-oriented form of psychotherapy that helps individuals identify and change negative thought patterns and behaviors. It is widely used to treat conditions such as depression, anxiety, and phobias by promoting healthier thinking and coping strategies.

In a randomized controlled trial, the effectiveness of Woebot's AI-guided chatbot in reducing depressive symptoms was comparable to clinician-led therapy, achieving similar clinical outcomes. Moreover, users formed a therapeutic bond with Woebot in three to five days, compared to two to six weeks with human therapists, and this bond remained stable over time without human interaction.[28] The cost-effectiveness of Woebot's AI-driven solution is expected to produce demonstrable improvement over traditional therapeutic models.

On the clinical side, AI-enhanced pharmacogenomics and NLP tools like Aifred Health are helping psychiatrists match patients to the right antidepressant. In a randomized controlled trial of 61 patients, Aifred's AI-driven Clinical Decision Support System achieved a 28.6% remission rate compared to none in the control group.[29] Further research demonstrated that patients treated with AI-guided prescribing utilizing pharmacogenomic data were 49% more likely to respond to medication and as much as 78% more likely to achieve remission, confirming the value of AI in the treatment of behavioral health.[30]

Chronic Disease

Finally, as we learned in Chapter 2, chronic diseases like diabetes, asthma, and hypertension drive the majority of healthcare spending and almost four times the mortality of all other causes combined.[31] Standard protocols often fail to consider individual variation in metabolism, behavior, or response to treatment.

Precision management, powered by AI and real-time data, is beginning to shift that. In one of the largest segments of chronic disease—diabetes—a notable example is Livongo (now part of Teladoc Health), which uses AI to interpret continuous glucose monitor (CGM) data, dietary inputs, and behavioral signals to personalize diabetes management.

In a 2017 outcomes analysis, Livongo patients saw a 0.9% reduction in HbA1c over one year. The program reported net cost savings of $83 per participant per month, or 5.8% compared to non-Livongo patients, mainly through reduced acute care utilization.[32] AI also helps detect nonadherence and behavioral patterns, nudging patients with timely prompts that improve engagement and outcomes.

Another example is Propeller Health's solution for asthma and COPD. Propeller uses connected inhaler sensors and AI-driven analytics to monitor medication usage, environmental triggers, and symptom patterns in real time. Instead of relying solely on patient self-reporting or sporadic clinical visits, the system personalizes care by detecting early warning signs of exacerbations and adjusting treatment prompts accordingly.

In clinical outcomes studies, Propeller users experienced a 57% reduction in asthma-related emergency department visits (35% reduction for COPD) and a 58% improvement in daily medication adherence.[33] These improvements translated to estimated healthcare savings of up to $2,475 per patient per year, largely through avoided hospitalizations and urgent care interventions.[34]

Omada Health is also advancing precision medicine in hypertension and cardiovascular risk management. Omada's platform combines connected blood pressure cuffs, lifestyle modification programs, AI-powered coaching, and remote clinical monitoring to tailor interventions for each participant. Unlike static educational materials or standard clinic visits, Omada dynamically adjusts care plans based on real-time vitals, engagement levels, and behavioral trends.

Participants enrolled in Omada programs have achieved average systolic blood pressure reductions of 8.1 mmHg and diastolic reductions of 4.1 mmHg after one year—gains that significantly lower the risk of cardiovascular events.[35] Importantly, these clinical improvements have produced estimated cost savings of $1,981 per member per year, primarily by reducing hospitalizations for hypertensive crises and cardiovascular complications.[36]

Caveats: Sometimes Precision Is More

The preceding examples notwithstanding, precision medicine is not always cheaper. In some cases, it can drive up costs, especially if testing is overused or treatment choices are not evidence-based. For instance, broad genomic testing offered without clear clinical indications can yield variants of uncertain significance, prompting further, often unnecessary, workups. Without robust AI decision support to contextualize risks and benefits, precision may become excess.

The lesson? Value arises when AI-enhanced precision is applied selectively, thoughtfully, and with clear benefit-to-cost alignment.

AI-assisted precision medicine is not a silver bullet, but when aimed at the right targets, it transforms medicine from a reactive, fragmented system into a personalized, proactive one. In doing so, it often improves outcomes and reduces costs.

From cancer to chronic disease, rare diagnoses to behavioral health, the evidence points to a powerful conclusion: precision isn't just about accuracy, it's about value. Systems that integrate AI, genomics, and real-world data are well-positioned to lead this transformation. The next frontier will be scaling these approaches equitably, so their benefits accrue not just to those who can afford them, but to every patient, everywhere.

Companies Shaping the Future of AI-Enabled Medicine

Beyond the established examples, the genomic revolution is rapidly translating theoretical promise into practical application. Pioneering companies are advancing care by integrating AI with genomic data to personalize treatments, improve outcomes, and reduce costs every day. The following are some of the key players advancing AI-enabled medicine.

Cancer Detection and Treatment

Grail. Grail's Galleri test employs AI to analyze freely circulating DNA fragments in the bloodstream, enabling the detection of more than 50 cancer types through a single blood draw. The test uses machine learning models trained on large genomic datasets to recognize subtle methylation signals associated with different cancers, distinguishing malignant from nonmalignant samples with high specificity. By integrating AI-driven pattern recognition with biological insights, Galleri is able to detect cancers often missed by traditional screening methods at earlier stages.[37]

Tempus. Tempus, mentioned earlier in the chapter, matches AI-driven tumor sequencing with precision treatment options. It collaborates with over 65% of U.S. academic medical centers and more than 50% of oncologists, facilitating personalized cancer care. Analyzing structured and unstructured clinical data alongside genomic profiles and matching targeted therapies, this AI-guided integration optimizes treatment selection and streamlines access to precision care.[38]

Foundation Medicine. Foundation Medicine has developed FDA-approved companion diagnostics that analyze more than 300 cancer-related genes to guide targeted therapies. FoundationOne tests use AI-driven algorithms to detect clinically significant mutations, interpret complex genomic signatures, and match patients to appropriate treatments. This process transforms raw sequencing data into patient-specific insights, helping clinicians make precise treatment decisions faster.[39]

Cardiometabolic and Preventive Health

Cleerly. Cleerly has developed an AI-powered coronary imaging platform that assesses heart disease risk by analyzing coronary computed tomography angiography (CCTA) scans. Its machine learning algorithms automatically identify, quantify, and characterize atherosclerotic plaques, providing detailed, objective assessments that traditionally required expert interpretation. By transforming complex imaging data into actionable insights, Cleerly's platform enables earlier and more precise detection of cardiovascular risk.[40]

Klinrisk. Klinrisk has developed an AI-powered model that predicts an individual's risk of chronic kidney disease (CKD) progression by analyzing patterns in laboratory tests, demographics, and clinical history. The Klinrisk model enables early identification of high-risk patients. The model has been validated in large U.S. datasets and is currently available for payers and health plans in population-level risk management, but it is not yet approved for clinical use in the United States.[41]

K Health. K Health and its similarly named app utilize AI to power a virtual primary care platform, analyzing millions of anonymized electronic health records and user-reported symptoms to generate personalized diagnostic insights. Its machine learning models identify probable conditions by comparing each user's case against a large dataset of similar medical histories, enabling providers to make faster, more data-informed decisions during remote consultations.[42]

Pharmacogenomics and Microbiome

Arine. Arine employs advanced AI algorithms to manage and optimize medication therapies across complex patient populations. The company's platform integrates pharmacogenomic insights, medication history, and clinical data to guide payers and providers regarding high-risk patients. By continuously learning from patient outcomes, Arine's AI enables smarter interventions that enhance adherence, minimize risks, and close critical gaps in medication management, especially for individuals with chronic conditions and polypharmacy.[43]

BiomeSense. BiomeSense combines continuous microbiome monitoring with AI-driven analytics to unlock new frontiers in precision health. Using its GutLab home sampling technology and MetaBiome bioinformatics platform, BiomeSense captures real-time microbiome data and applies machine learning models to uncover actionable microbial patterns. This enables more precise predictions of health status and guides targeted interventions based on individual gut microbiome signatures.[44]

Rare Disease and Pediatric Genomics

Fabric Genomics. Fabric Genomics' Fabric GEM, an AI-powered analytic platform, accelerates the interpretation of whole genomes and exomes. Using advanced machine learning algorithms, Fabric GEM analyzes millions of genetic variants, prioritizes likely disease-causing mutations, and generates structured reports to support clinical decision-making. By automating complex genomic analysis, the platform enables faster, more scalable precision diagnostics across a wide range of rare and inherited diseases.[45]

Face2Gene. As highlighted in the "Cost and Outcomes" section, FDNA has developed Face2Gene, an AI-powered platform that analyzes facial features to assist in the diagnosis of rare genetic disorders. Using deep learning algorithms trained on large datasets of clinical images, Face2Gene identifies subtle phenotypic patterns that indicate underlying genetic conditions. By augmenting traditional diagnostic methods, the platform supports faster and more accurate genetic evaluations.[46]

Rady Children's Institute for Genomic Medicine. Rady Children's Institute has developed an AI-assisted pipeline that automates the analysis and interpretation of pediatric whole-genome sequencing data. By integrating machine learning algorithms with clinical databases and variant prioritization tools, the platform rapidly identifies potential disease-causing mutations in critically ill infants.[47]

These advances are outstanding, but innovation driven by pioneering companies is only part of the story. As precision medicine moves from research institutions into everyday care, its future increasingly depends on personal involvement. Patients, participants, and advocates now have the power to shape the next generation of healthcare by choosing, supporting, and engaging with these new tools and approaches. Let's look next at how this can be accelerated at the user level.

What You Can Do

Although many of the technologies discussed may seem cutting-edge or confined to research settings, there are increasingly practical steps individuals can take to engage with precision medicine today:

- *Get a wearable device.* Use digital health tools such as Apple Watch or Fitbit that integrate with platforms like Livongo. These tools help personalize care and generate the kind of real-world data AI systems need to improve recommendations.

- *Talk to your doctor.* Ask your doctor about pharmacogenomic testing if you've experienced multiple medication side effects or have a complex treatment history. Tests are often covered by insurance for behavioral health, cardiac, and other conditions.

- *Choose precision medicine providers.* Select care teams or health systems that advertise precision medicine integration, whether through genetic counseling, AI-powered screening, or access to genomic sequencing.

- *Stay informed.* Connect with trusted resources like the Mayo Clinic, NIH, or your local academic medical center. Many now offer free webinars or consultations on genetic and personalized care options.

- *Join research studies.* Large-scale studies like the NIH's All of Us Research Program and your health system's local biobank contribute to the development of more inclusive and powerful AI models, ensuring that future tools reflect diverse populations.

- *Advocate for better access.* If you or a loved one would benefit from genomic testing or precision approaches, understanding insurance criteria and working with providers can often overcome bureaucratic barriers.

These are not just passive decisions. As patients, participants, and advocates, individuals help drive the adoption of smarter, more personalized, and more equitable care.

The shift from population-level medicine to biologically individualized care represents one of the most profound changes in healthcare since the advent of antibiotics or vaccines. Enabled by AI, powered by genomics, and delivered through both startups and world-class institutions, precision medicine is no longer a concept for tomorrow—it's an evolving standard for today. Yet its pace and reach depend on more than innovation. It requires infrastructure, education, and engagement.

As we move into Chapter 12, we'll explore what lies beyond today's breakthroughs: how emerging models of care, data sharing, trust, and AI co-evolution will shape the next generation of medicine. The future isn't just something to wait for—it's something to help build.

Notes

1. Colby, "Whole Genome Sequencing Cost."

2. National Institutes of Health, "DNA Terrain Affects Function in Human Genome," March 23 (2009), https://www.nih.gov/news-events/nih-research-matters/dna-terrain-affects-function-human-genome.

3. UK Biobank, "AI-based Integrative Risk Scores for Predictive Medicine and Precision Therapy in Complex and Rare Diseases," January 19 (2023), https://www.ukbiobank.ac.uk/enable-your-research/approved-research/ai-based-integrative-risk-scores-for-predictive-medicine-and-precision-therapy-in-complex-and-rare-diseases.

   U.S. Department of Veterans Affairs, "Million Veteran Program (MVP)," November 15 (2023), https://www.research.va.gov/mvp.

4. Tempus, "Expanding the Boundaries of Precision Medicine and Research with Tempus One," August 17 (2023), https://www.tempus.com/resources/content/blog/expanding-the-boundaries-of-precision-medicine-and-research-with-tempus-one.

5. Deodhar, Malavika, Rihani, Sweilem, Darakjian, Lucy, Turgeon, Jacques, and Michaud, Veronique, "Assessing the Mechanism of Fluoxetine-Mediated CYP2D6 Inhibition," *Pharmaceutics* 13, no. 2 (2021): 148.

6. U.S. Food and Drug Administration, "The Software Precertification (Pre-Cert) Pilot Program: Tailored Total Product Lifecycle Approaches and Key Findings," September (2022), https://www.fda.gov/media/161815/download.

7. Watson, Michael, Lloyd-Puryear, Michele, and Howell, Rodney, "The Progress and Future of US Newborn Screening," *International Journal of Neonatal Screening* 8, no. 3 (2022): 41.

8. Hughes, Julia, Shymka, Mikayla, Ng, Trevor, Phulka, Jobanjit, Safabakhsh, Sina, and Laksman, Zachary, "Polygenic Risk Score Implementation into Clinical Practice for Primary Prevention of Cardiometabolic Disease," *Genes (Basel)* 15, no. 12 (2024): 1581.

9. Buchanan, Bruce, and Shortliffe, Edward, *"Rule-Based Expert Systems: The MYCIN Experiments of the Stanford Heuristic Programming Project,"* Addison-Wesley (1984).

10. Norvig, Peter, *"Paradigms of Artificial Intelligence Programming: Case Studies in Common Lisp,"* Morgan Kaufmann (1992).

11. Campbell, Dennis, "UK Medicines Regulator Approves Gene Therapy for Two Blood Disorders," November 16 (2023), https://www.theguardian.com/society/2023/nov/16/uk-medicines-regulator-approves-casgevy-gene-therapy-for-two-blood-disorders-sickle-cell.

12. Guess, Angela, "IBM and Quest Diagnostics Launch Watson-Powered Genomic Sequencing Service," October 24 (2016), https://www.dataversity.net/ibm-quest-diagnostics-launch-watson-powered-genomic-sequencing-service.

13. Arrieta, Oscar, Catalán, Rodrigo, Guzmán-Vazquez, Silvia, Barrón, Feliciano, Lara-Mejía, Luis, Soto-Molina, Herman, Ramos-Ramírez, Maritza, Flores-Estrada, Diana, and de la Garza, Jaime, "Cost-effectiveness Analysis of First and Second-Generation EGFR Tyrosine Kinase Inhibitors as First Line of Treatment for Patients with NSCLC Harboring EGFR Mutations," *BMC Cancer* 20 (2020): 829.

14. Khera, Amit, Chaffin, Mark, Aragam, Krishna, Haas, Mary, Roselli, Carolina, Hoan Choi, Seung, Natarajan, Pradeep, Ellinor, Patrick, Lander, Eric, Lubitz, Steven, and Kathiresan, Sekar, "Genome-wide Polygenic Scores for Common Diseases Identify Individuals with Risk Equivalent to Monogenic Mutations," *Nature Genetics* 50, no. 9 (2018): 1219–1224.

15. Inside Precision Medicine, "Intermountain Study Finds Active Approach to Statin Selection Improves Outcome," March 6 (2023), https://www.insideprecisionmedicine.com/news-and-features/ intermountain-study-finds-active-approach-to-statin-selection-improves-outcome.

16. O'sullivan, Jack, Raghavan, Sridharan, Marquez-Luna, Carla, Luzum, Jasmine, Damrauer, Scott, Ashley, Euan, O'Donnell, Christopher, Willer, Cristen, and Natarajan, Pradeep, "Polygenic Risk Scores for Cardiovascular Disease: A Scientific Statement From the American Heart Association," *Circulation* 146, no. 8 (2022): e93–e118.

Samani, Nilesh, Beeston, Emma, Greengrass, Chris, Riveros-McKay, Fernando, Debiec, Radoslaw, Lawday, Daniel, Wang, Qingning, Budgeon, Charley, Braund, Peter, Bramley, Richard, Kharodia, Shireen, Newton, Michelle, Marshall, Andrea, Krzeminski, Andre, Zafar, Azhar, Chahal, Anuj, Heer, Amadeeep, Khunti, Kamlesh, Joshi, Nitin, Lakhani, Mayur, Farooqi, Azhar, Plagnol, Vincent, Donnelly, Peter, Weale, Michael, and Nelson, Christopher, "Polygenic Risk Score Adds to a Clinical Risk Score in the Prediction of Cardiovascular Disease in a Clinical Setting," *European Heart Journal* 45, no. 34 (2024): 3152–3160.

Khera, Rohan, Pandey, Ambarish, Ayers, Colby, Carnethon, Mercedes, Greenland, Philip, Ndumele, Chiadi, Nambi, Vijay, Seliger, Stephen, Chaves, Paulo, Safford, Monika, Cushman, Mary, Xanthakis, Vanessa, Vasan, Ramachandran, Mentz, Robert, Correa, Adolfo, Lloyd-Jones, Donald, Berry, Jarrett, de Lemos, James, and Neeland, Ian, "Performance of the Pooled Cohort Equations to Estimate Atherosclerotic Cardiovascular Disease Risk by Body Mass Index," *JAMA Network Open* 3, no. 10 (2020): e2023242.

Silva, Matthew, Swanson, Anna, Gandhi, Pritesh, and Tataronis, Gary, "Statin-related Adverse Events: A Meta-analysis," *Clinical Therapeutics* 28, no.1 (2006): 26-35.

Ashman, Jill, Santo, Loredana, and Okeyode, Titilayo, "Characteristics of Office-based Physician Visits by Age, 2019," *National Health Statistics Reports* 184 (2023).

Sihvonen, Marja, and Kekki, Pertti, "Unnecessary Visits to Health Centres as Perceived by the Staff," *Scandinavian Journal of Primary Health Care* 8, no. 4 (1990): 233-239.

Owens, Pamela, Barrett, Marguerite, and Stocks, Carol, "Characteristics and Costs of Potentially Preventable Inpatient Stays, 2017," June (2020), https://hcup-us.ahrq.gov/reports/statbriefs/sb259-Potentially-Preventable-Hospitalizations-2017.pdf.

Slight, Sarah, Seger, Diane, Franz, Calvin, Wong, Adrian, and Bates, David, "The National Cost of Adverse Drug Events Resulting from Inappropriate Medication-related Alert Overrides in the United States," *Journal of the American Medical Informatics Association* 25, no. 9 (2018): 1183-1188.

Hamblin, Susan, Rumbaugh, Kelli, and Miller, Richard, "Prevention of Adverse Drug Events and Cost Savings Associated with PharmD Interventions in an Academic Level I Trauma Center: An Evidence-based Approach," *Journal of Trauma and Acute Care Surgery* 73, no.6 (2012): 1484-1490.

Fay, Bill, "How Much Will a Doctor Visit Cost You?," December 11 (2024), https://www.debt.org/medical/doctor-visit-costs.

National Center for Health Statistics, "Hospitalization – Health, United States," June (2023), https://www.cdc.gov/nchs/hus/topics/hospitalization.htm.

17. Van Driest, Sara, Shi, Yaping, Bowton, Erica, Schildcrout, Jonathan, Peterson, Josh, Pulley, Jill, Denny, Joshua, and Roden, Dan, "Clinically Actionable Genotypes Among 10,000 Patients with Preemptive Pharmacogenomic Testing," *Clinical Pharmacology and Therapeutics* 95, no. 4 (2014): 423-431.

18. Krebs, Kristi, and Milani, Lili, "Translating Pharmacogenomics into Clinical Decisions: Do Not Let the Perfect be the Enemy of the Good," *Human Genomics* 13, no. 39 (2019).

19. Mayo Clinic, "Pharmacogenomics Program – Center for Individualized Medicine," April (2025), https://www.mayo.edu/research/centers-programs/center-individualized-medicine/research/pillars-programs/pharmacogenomics.

20. David, Victoria, Fylan, Beth, Bryant, Eleanor, Smith, Heather, Sagoo, Gurdeep, and Rattray, Marcus, "An Analysis of Pharmacogenomic-Guided Pathways and Their Effect on Medication Changes and Hospital Admissions: A Systematic Review and Meta-Analysis," *Frontiers in Genetics* 12 (2021): 698148.

21. Ibid.

22. Miotto, Riccardo, Li, Li, Kidd, Brian, and Dudley, Joel, "Deep Patient: An Unsupervised Representation to Predict the Future of Patients from the Electronic Health Records," *Scientific Reports* 6 (2016): 26094.

23. Fox, Andrea, "Mount Sinai Health Announces New Center for AI and Human Health," November 25 (2024), https://www.health careitnews.com/news/mount-sinai-health-announces-new-center-ai-and-human-health.

24. Palantir Technologies, "Cleveland Clinic / Palantir," April (2025), https://www.palantir.com/impact/cleveland-clinic.

25. EveryLife Foundation for Rare Diseases, "The Cost of Delayed Diagnosis in Rare Disease: A Health Economic Study," April (2025), https://everylifefoundation.org/delayed-diagnosis-study.

26. Yahya, Dinnar, Stoyanova, Milena, Hachmeriyan, Mari, and Levkova, Mariya, "The Application of the Facial Analysis Program Face2Gene in a Single Genetic Counseling Center: A Retrospective Study," *Egypt Pediatric Association Gazette* 73, no. 1 (2025).

27. Undiagnosed Diseases Network, "Undiagnosed Diseases Network," April (2025), https://undiagnosed.hms.harvard.edu.

28. Darcy, Alison, Daniels, Jade, Salinger, David, Wicks, Paul, and Robinson, Athena, "Evidence of Human-Level Bonds Established With a Digital Conversational Agent: Cross-sectional, Retrospective Observational Study," *JMIR Formative Research* 5, no. 5 (2021): e27868.

29. Benrimoh, David, Whitmore, Kate, Richard, Maud, Golden, Grace, Perlman, Kelly, Jalali, Sara, Friesen, Timothy, Barkat, Youcef, Mehltretter, Joseph, Fratila, Robert, Armstrong, Caitrin, Israel, Sonia, Popescu, Christina, Karp, Jordan, Parikh, Sagar, Golchi, Shirin, Moody, Erica, Shen, Junwei, Gifuni, Anthony, Ferrari, Manuela, Sapra, Mamta, Kloiber, Stefan, Pinard, Georges, Dunlop, Boadie, Looper, Karl, Ranganathan, Mohini, Enault, Martin, Beaulieu, Serge, Rej, Soham, Hersson-Edery, Fanny, Steiner, Warren, Anacleto, Alexandra, Qassim, Sabrina, McGuire-Snieckus, Rebecca, and Margolese, Howard, "Artificial Intelligence in Depression – Medication Enhancement (AID-ME): A Cluster Randomized Trial of a Deep Learning Enabled Clinical Decision Support System for Personalized Depression Treatment Selection and Management," *medRxiv* (2024), https://www.medrxiv.org/content/10.1101/2024.06.13.24308884v1.

30. Tesfamicael, Kiflu, Zhao, Lijun, Fernández-Rodríguez, Rubén, Adelson, David, Musker, Michael, Polasek, Thomas, Lewis, Martin, "Efficacy and Safety of Pharmacogenomic-Guided Antidepressant Prescribing in Patients with Depression: An Umbrella Review and Updated Meta-analysis," *Frontiers in Psychiatry* 15 (2024): 1276410.

31. Hajat and Stein, "The Global Burden of Multiple Chronic Conditions."

32. Livongo, "Livongo for Diabetes: Clinical Outcomes and Cost Savings Case Study," April (2025), https://content.teladochealth.com/Announcements/LivongoCaseStudy-costsavings-RBG-WEB(1).pdf.

33. Digital Therapeutics Alliance, "Propeller," April (2025), https://dtxalliance.org/products/propeller.

34. Inocencio, Timothy, Sterling, Kimberly, Sayiner, Sibel, Minshall, Michael, Kaye, Leanne, and Hatipoğlu, Umur, "Budget Impact Analysis of a Digital Monitoring Platform for COPD," *Cost Effectiveness and Resource Allocation* 21, no. 1 (2023): 36.

35. Wu, Justin, Napoleone, Jenna, Linke, Sarah, Noble, Madison, Turken, Michael, Rakotz, Michael, Kirley, Kate, Folk Akers, Jennie, Juusola, Jessie, and Jasik, Carolyn, "Long-Term Results of a Digital Hypertension Self-Management Program: Retrospective Cohort Study," *JMIR Cardio* 7 (2023): e43489.

36. American Medical Association, "Future of Health Case Study: Omada Health," April (2025), https://www.ama-assn.org/system/files/future-health-case-study-omada-health.pdf.

37. GRAIL, "Final Results from PATHFINDER Study of GRAIL's Multi-Cancer Early Detection Blood Test Published in The Lancet," October 5 (2023), https://grail.com/press-releases/final-results-from-pathfinder-study-of-grails-multi-cancer-early-detection-blood-test-published-in-the-lancet.

38. Tempus, "Tempus: AI-enabled Precision Medicine," April (2025), https://www.tempus.com.

39. Foundation Medicine, "Foundation Medicine Receives FDA Approval for FoundationOne® CDx and FoundationOne® Liquid CDx as Companion Diagnostics for LYNPARZA® in BRCA-Mutated Metastatic Castration-Resistant Prostate Cancer," September 3 (2024), https://www.foundationmedicine.com/press-release/fda-approval-cdxs-lynparza-brca-mcrpc.

40. Cleerly Health, "Research on AI-Enabled Quantitative CT Coronary Assessment," October 28 (2024), `https://cleerlyhealth.com/press/research-on-ai-enabled-quantitative-ct-coronary-assessment`.

41. Klinrisk, "Klinrisk: AI-Powered Risk Prediction for Chronic Kidney Disease," April (2025), `https://www.klinrisk.com`.

42. K Health, "K Health: Smarter, Personalized Primary Care," April (2025), `https://khealth.com`.

43. Arine, "Arine: AI-Driven Precision Medication Management," April (2025), `https://www.arine.io`.

44. BiomeSense, "BiomeSense: Real-Time Microbiome Monitoring and Analytics," April (2025), `https://www.biomesense.com`.

45. Fabric Genomics, "Fabric Genomics: AI-Driven Genomic Interpretation for Precision Medicine," April (2025), `https://fabricgenomics.com`.

46. FDNA, "Face2Gene: AI-Driven Facial Analysis for Genetic Diagnostics," April (2025), `https://www.face2gene.com`.

47. Rady Children's Institute for Genomic Medicine, "Rady Genomics: Advancing Rapid Whole Genome Sequencing for Critically Ill Children," April (2025), `https://radygenomics.org`.

A Vision for the Future

Having explored the foundations, real-world applications, and challenges of precision medicine, we now turn to what comes next. In this chapter, we look at the emerging capabilities that are beginning to reshape how healthcare is delivered—faster, smarter, and more personally than ever before.

Precision medicine is entering a period of rapid convergence, where technologies that once operated in silos—genomic sequencing, artificial intelligence (AI), mobile health, and clinical data platforms—are beginning to interlock into actionable, patient-specific systems. Instead of isolated applications or one-off pilot programs, we're seeing integrated ecosystems emerge that update in real time, adapt to individual patients, and scale across populations.

At the heart of this convergence is the growing availability of interoperable data and standardized infrastructure. In the next several years, health systems will increasingly adopt dynamic care engines that personalize treatment recommendations based on evolving input from biometric devices, EHRs, genetic profiles, and behavioral signals. These dynamic, integrated, technology-enabled systems will simultaneously synthesize robust, real-time information at the individual level and compare it to population-level relationships between treatments and outcomes previously unavailable to the clinician, revolutionizing care delivery.

The future is coming fast—let's take a look.

What's Next for Precision Medicine

An interesting early-stage example of dynamic, technology-enabled treatment is the creation of "digital twins" in healthcare—virtual representations of patients that simulate treatment responses based on real-world and biological data. Although previously limited to experimental environments, these twins are now being used by companies like Unlearn.AI in clinical trials and beginning to appear in early-care personalization strategies.[1] Similarly, platforms like Tempus, mentioned previously (Chapter 11), have moved from niche genomic applications to become foundational engines in clinical information and insights.[2]

These shifts are not speculative. They reflect an active migration of tools from academic labs into provider workflows and payer decision-making. As care platforms become increasingly modular and AI-native, we can expect the clinical experience to move from fixed pathways to flexible, predictive ones that continuously learn from patient outcomes.

The Four Drivers

Four key areas—our familiar drivers of precision medicine—highlight the technologies and capabilities now leading the transition from static protocols to adaptive, patient-specific care.

Next-Generation Wearables

Wearables are already evolving beyond consumer-grade wellness tools into clinically validated health monitors capable of driving therapeutic decisions. The next generation of devices will continue this path, incorporating multisensor arrays to monitor not just heart rate and sleep but also key biomarkers like cortisol (stress), lactate (metabolism), and oxygen saturation with medical-grade precision.

These tools will blur the line between diagnostics and everyday wellness support, forming a "daily health radar" that delivers early warnings for acute and chronic conditions alike. Crucially, these

> Tools will blur the line between diagnostics and wellness support, forming a "daily health radar" that delivers early warnings for acute and chronic conditions.

systems won't just alert patients—they'll suggest actionable next steps and notify care teams when necessary, with contextual AI guidance replacing generalized nudges.

Consider Abbott's Lingo devices, which began in the health-aware non-diabetes market but are now moving into adjacent use cases, including metabolic optimization.[3] Similarly, Apple's HealthKit ecosystem and expanded sleep apnea, blood pressure monitoring, AI doctor, and Apple Health Study functionality signal the transition toward advanced digital therapeutics.[4]

In the near term, it's likely that these devices and other ambient coaching platforms like Whoop and Ultrahuman will connect with integrators such as Validic (an electronic healthcare solution for remote patient care) to facilitate sleep hygiene, nutrition, and fitness in pursuit of precision preventive care.[5] These systems in turn will increasingly integrate with clinical platforms such as Epic and Cerner, allowing clinicians to monitor patient progress remotely and make adjustments on the fly.

Preemptive Pharmacogenomics

Pharmacogenomics is poised to become a routine part of care pathways rather than a specialized outlier. With testing costs approaching the double-digit price point and turnaround times shrinking to under a week, providers are increasingly ordering pharmacogenomic panels preemptively rather than reactively. The use of combinatorial PGx—assessing how multiple genetic markers interact to influence drug metabolism—adds nuance and clinical utility.

This isn't just a theoretical benefit. Mayo Clinic's RIGHT 10K program and the Department of Veterans Affairs' PHASER program have already demonstrated that preemptive pharmacogenomic testing reduces adverse drug reactions and improves medication efficacy.[6] Companies like Genomind and Gene By Gene offer clinical-grade PGx panels for psychiatric and primary care implementation, and payer coverage is expanding to match clinical utility.[7]

The downstream impact of preemptive use will be profound: fewer failed prescriptions, reduced trial-and-error cycles, improved adherence, and lower total costs. As pharmacogenomic data is embedded in a patient's longitudinal record, medication reconciliation and deprescribing efforts will also benefit, enabling more effective polypharmacy management in elderly and complex patients.

Big Data and Predictive Population Health

With the exponential growth of data sources—from electronic health records (EHRs) to wearable outputs to claims histories—the shift from reactive to predictive care is not only possible but inevitable. Machine learning models can now parse patterns across millions of data points to flag health deterioration days or even weeks before clinical symptoms appear. Within the next few years, healthcare systems will rely increasingly on predictive algorithms to allocate care management resources dynamically, identifying high-risk patients who may otherwise go unnoticed until a hospitalization or emergency event.

Early examples of these models are already in use. Kaiser Permanente's sepsis prediction algorithm reduces ICU mortality by flagging deteriorating patients hours before traditional vital-sign-based methods (see Figure 12-1).

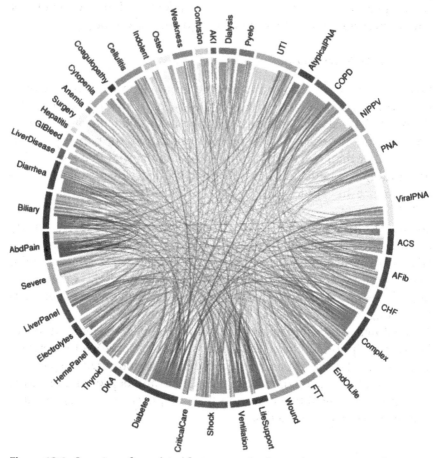

Figure 12-1: Overview of correlated factors contributing to the occurrence of sepsis

Source: [8]/Kaiser Permanente Division of Research

Similar tools are being developed to anticipate hospital readmission, detect medication non-adherence, and forecast disease flares in chronic conditions like Crohn's and asthma. Practical research at Purdue University and the University of Edinburgh are but two examples of early success in this regard.[9] The challenge moving forward will be to translate these predictions into real-time interventions. This will require coordination between AI platforms, clinical workflows, and frontline staff—a shift from retrospective data audits to embedded, anticipatory care strategies.

AI and Human Interfaces

The next wave of AI in medicine will offer not only answers, but explanations—becoming collaborators rather than calculators. In high-stakes medicine, clinicians demand interpretability: they must understand why an AI recommends a certain treatment or flags a risk. In the next few years, advances in explainable artificial intelligence (XAI) will enable systems to provide evidence trails, risk comparisons, and citations from current literature. This will make AI more acceptable and trusted, especially in domains like oncology, cardiology, and neurology.

For example, clinical assistants—large language models (LLMs) fine-tuned on medical corpora and augmented with clinical AI systems—are accelerating implementation in diagnosis, documentation, and patient communication. Microsoft's DAX Copilot, developed with Nuance, is already integrating with systems like Epic and Cerner, turning real-time conversations between doctors and patients into structured clinical notes.[10] Today it's helping reduce time spent on paperwork; tomorrow, it may evolve into a true digital copilot—one that listens, learns, and supports medical decision-making as care progresses.

Although the time spent on different activities of physicians may change, like the calculator (and later the computer) was to the slide rule for engineers, these systems won't replace physicians; they'll enhance decision-making under time constraints and reduce cognitive load. A genomic variant flagged in a cancer report, for instance, could trigger AI-generated summaries of relevant trials or evidence-based off-label options, making precision insights both rapid and accessible. Some of the most advanced uses are already emerging from companies like PathAI in pathology and Aidoc in radiology, which integrate into clinical workflows and support multimodal information (e.g., imaging, genomics, labs).[11]

Emerging technologies like Neuralink, which translate cognitive intent directly into motor commands, represent the ultimate frontier in human–computer interaction.

On the horizon, emerging technologies like Neuralink, which aims to translate cognitive intent directly into digital or motor commands, represent the ultimate frontier in human–computer interaction.[12] Although still experimental, such brain–computer interfaces could one day extend the boundaries of personalized healthcare—particularly for individuals with severe motor impairments—by restoring communication or enabling control of assistive devices through thought alone.

Cross-Sector Integration and Ecosystem Platforms

Precision medicine depends on precision data, but that data has long been fragmented across labs, insurers, EHR vendors, and point-of-care tools. What's changing now is the emergence of commercially viable platforms that bridge these silos.

Interoperability standards like Fast Healthcare Interoperability Resources (FHIR), alongside cloud-native infrastructures, are enabling more seamless sharing of genetic, biometric, and clinical information.[13] With initiatives like the Trusted Exchange Framework and Common Agreement (TEFCA), we will see a dramatic expansion of "shared-care ecosystems" where multiple stakeholders contribute to and benefit from a unified patient profile.[14]

One pioneering example is PointClickCare's integration of EHR, pharmacy, and remote monitoring data for post-acute skilled nursing facilities and long-term-care populations.[15] Others include Epic's Cosmos platform and AWS HealthLake, which aim to create longitudinal health records across settings.[16] As these integrations continue to gain adoption, they will accelerate smarter prior authorization, more precise risk scoring, and automated care navigation across complex health journeys.

As genomic data is increasingly layered into these environments, we'll also see more precise clinical alerts, better medication reconciliation, and less reliance on static protocols. This isn't just a technological feat—it's a shift in the operating system of healthcare itself.

From Episodic to Continuous Care

Healthcare delivery is being fundamentally restructured by the availability of real-time, patient-generated data. Rather than episodic visits every 6–12 months, patients can engage in continuous care relationships through passive monitoring, digital nudges, and asynchronous check-ins. This profoundly changes the interaction between provider and patient. For clinicians, this means gaining access to detailed health trajectories, not just snapshots. For patients, it means the ability to receive timely adjustments and interventions without needing to initiate contact or travel to a clinic. The result is more personalized, more convenient, and more effective care.

A good example is Omada Health, which offers behavior-change coaching for diabetes, hypertension, and joint and muscle health via a mobile app, integrated with remote monitoring tools and health coaches.[17] Another is Current Health, whose wearable and platform are used to manage hospital-at-home programs with continuous vitals tracking.[18]

In the next several years, we can expect a proliferation of hybrid care models where a wearable alerts a care coordinator, who texts the patient and arranges a virtual consult—creating a seamless chain of responsiveness. These systems will transform both the quality and the cadence of care, moving patients to proactive alignment rather than reactive crisis.

Implications for Traditional Medicine

Traditional medicine has long relied on a standardized, population-level model of care, a framework that assumes what works for the average patient will suffice for most. Although this approach enabled the creation of protocols, formularies, and payer models at scale, it increasingly struggles to serve patients whose needs deviate from the mean. The result is often a disjointed system in which patients cycle through specialists, repeat diagnostic tests, and trial therapies with mixed results and mounting frustration.

The rigidity of the traditional model of care is under pressure from multiple directions. Clinicians face mounting documentation burdens and productivity quotas that crowd out personalized attention. Patients feel reduced to checkboxes. Meanwhile, the system's cost curve continues to rise, driven by avoidable adverse events, overprescription, and the failure to detect high-risk individuals early. These issues are not merely operational. They stem from the foundational misalignment between the heterogeneity of biology and the generality of traditional care.

Precision medicine doesn't eliminate this pressure; it redirects it. As more data becomes available at the individual level, the expectation grows that care will adapt in kind. But traditional models are not designed to do so. Without structural change, financially, technologically, and philosophically, healthcare will continue to underperform for those who need it most. What's needed is not just better medicine but a better system: one designed from the start to recognize difference, respond to complexity, and evolve in real time.

Clinical Workflow Innovation

The rise of precision medicine challenges the deeply entrenched workflows that define modern clinical practice. Traditional workflows follow a time-bound, visit-centered cadence, shaped largely by billing codes and risk-adjusted contracts, not necessarily by patient biology. A typical physician has 15–20 minutes per encounter to gather a history, make a diagnosis, and offer treatment, often relying on pattern recognition, heuristics, and past experience. That's a poor match for a system increasingly driven by complex data streams and individualized insights.

Precision medicine offers new inputs from wearable devices, genetic results, and AI-driven decision support. Traditional workflows are not built to absorb or act on them. Clinical systems lag in presenting this information in meaningful ways. Often, genomics reports and pharmacogenomic flags arrive days after a prescription is written. Even when available, physicians may lack the time, training, or confidence to interpret them. The result: missed opportunities to intervene early, tailor care, and avoid harm.

> The challenge is not to demand more from physicians but to redesign the role itself.

The challenge is not to demand more from physicians but to redesign the role itself. We must move from a model of solo heroic generalists to teams where information stewards, AI navigators, and clinical interpreters play an active role. Workflow innovation must be guided by human factors design and reimagined from the patient forward to the integrated, longitudinal care plan—not from the EHR and reimbursement rules backward to the patient. This isn't about adding new responsibilities but rather about freeing clinicians to practice at the top of their license, with tools that extend rather than constrict their capacity for care.

The effort to accomplish this begins with retooling medical education. Today's medical school curriculum still largely reflects a 20th-century model: organ systems taught in isolation, lectures anchored in population norms, and diagnostic reasoning focused on static case presentations. Although this foundation is essential, it fails to equip future clinicians to manage dynamic, data-driven, and individualized care. Medical students graduate knowing the Krebs cycle but not how to interpret a pharmacogenomic profile, interrogate a predictive algorithm, or navigate multimodal streams of patient-generated data.

NOTE   The Krebs cycle is the chemical process by which cells turn food into energy, using oxygen to produce fuel for the body and releasing carbon dioxide as waste.

The pace of discovery has simply outstripped the structure of education. Genomic medicine, digital therapeutics, AI-based diagnostics, and behaviorally guided interventions require new fluencies. Yet curricula change slowly. Precision medicine remains a "special topic" at most institutions, not a core competency. Worse, training often reinforces the utility-of-the-mean thinking that precision medicine explicitly challenges, producing clinicians who are comfortable treating syndromes but not people.

> New clinical roles may emerge—*forologists* who identify early-stage risk and guide health optimization before infirmity, *acutologists* who initiate and manage treatment at the onset of a condition, and *chronicists* and *cognicists* for chronic disease and behavioral health, respectively.

The future demands a rethinking of what it means to be clinically aligned with longitudinal, precision care. New roles may emerge—*forologists* who identify early-stage risk and guide health optimization before infirmity, *acutologists* who initiate and manage treatment at the onset of a condition, and *chronicists* and *cognicists* for chronic disease and behavioral health, respectively. But even traditional roles will need to evolve. Residency programs must integrate decision science, ethics of data use, and cross-disciplinary problem-solving. Lifelong learning must be embedded through modular micro-certifications and digital simulations. We must train not just doctors but adaptive learners prepared for a healthcare system that is no longer linear but layered, fluid, and precision-guided.

Redefining Evidence-Based Medicine

Evidence-based medicine (EBM) has been a guiding force in modern healthcare, bringing rigor and consistency to decision-making. But its reliance on randomized controlled trials (RCTs) and population-level outcomes, although statistically elegant, increasingly clashes with the reality of individualized care. The challenge lies in the core assumption of EBM: that what works best for the majority is a reliable proxy for what works best for you. In the age of precision medicine, that assumption is uncomfortably wrong.

RCTs are slow and expensive and often exclude patients with comorbidities, genetic variation, or social complexity—the very features that define real-world care. More concerning, these trials rarely generate insights about subpopulations or individuals with rare variants. Regulatory approval pathways and payer coverage often still depend on these average-based results, leaving precision-guided interventions with limited pathways to adoption. The result is a system in which innovation must first prove its worth in generalized terms before being allowed to demonstrate its strength in individualized practice.

A redefinition is underway. Real-world evidence (RWE), n-of-1 trials, adaptive designs, and AI-generated simulations are beginning to supplement, and sometimes supplant, traditional RCTs. Regulatory bodies like the FDA and EMA are acknowledging these new models, but clinical cultures are often slower to adapt. The future of EBM must retain its rigor while embracing its evolution: not abandoning generalizability, but pairing it with personalization. It must become evidence-based and individual-aware, capable of balancing statistical confidence with personal relevance.

Traditional medicine also pays for care within a finely tuned economic model that incentivizes volume over value, and uniformity over uniqueness. Most reimbursement frameworks are designed around procedural codes, diagnostic buckets, and average treatment pathways. Precision medicine, by nature more granular and often preventive, struggles to fit into this mold. As a result, many of its most promising tools, from pharmacogenomics to AI-guided decision support, remain underutilized or unreimbursed, even when cost-saving in the long run.

This defines the dilemma for regulators and payers. Precision approaches often lack the large-scale RCT data used to justify coverage decisions, yet they may also reduce future claims (by preventing adverse events or unnecessary interventions) while generating billable encounters today. This creates a perverse disincentive. Systems are penalized for improving care in ways that don't neatly align the timing of the problem with revenue. Meanwhile, providers navigating narrow formularies or prior authorization rules find it easier to prescribe what's covered for everyone, rather than what's best for someone.

The future will require new economic thinking. Value-based care contracts, risk-sharing arrangements, and longitudinal outcome tracking can support the adoption of precision tools that reduce the total cost of care. Regulators, too, must evolve, creating faster, more flexible approval pathways for algorithms, dynamic diagnostics, and personalized therapies. The shift won't be easy, but it is essential. If medicine is to become truly personalized, its payment and policy infrastructure must evolve from managing volume to enabling precision.

Hospital and Clinic Transformation

Traditional brick-and-mortar healthcare facilities were designed for an era when care was episodic, technology scarce, and diagnosis dependent on face-to-face examination. Although still vital for acute and procedural care, these structures are increasingly mismatched to the needs of a digital, decentralized, and data-enabled health system. The rise of precision medicine calls for environments that are flexible, distributed, and integrated with the rhythms of everyday life, not siloed in campuses built for an earlier age.

Hospitals continue to invest in expanding surgical centers and emergency wings while underinvesting in infrastructure that enables precision care—things like genomic labs, remote monitoring hubs, AI analytics centers, and virtual-first specialty consults. Clinics, similarly, remain geared toward in-person assessments of symptoms rather than interpretation of predictive data. The result is an experience that is reactive, fragmented, and costly.

Transformation is not about replacement but repurposing. In the near future, health systems will need precision pods: hybrid spaces where data is interpreted, care is coordinated, and virtual and physical care are blended seamlessly. Home-based care will expand, supported by wearable telemetry and remote teams. Ambulatory genomics centers may supplement traditional primary care offices. Facilities will need to accommodate not just patients but ecosystems of clinicians, informaticians, and AI systems collaborating in real time. Precision medicine will not abolish the hospital, but it will demand its reinvention.

Related to this is a reorientation of trust in the care delivery system. Trust has always been the currency of healthcare, but traditional systems have devalued it through opacity, inefficiency, and impersonality. Patients too often feel processed rather than cared for, with brief visits, unexplained decisions, and a lack of follow-through. As precision medicine introduces new layers of complexity, from genetic risk scoring to AI-assisted diagnosis, patient-centeredness must be reasserted as a core principle. Traditional models have failed to address this. Physicians overwhelmed by documentation demands have less time for empathy. Fragmented data and inconsistent records breed confusion. The paternalistic remnants of 20th-century medicine still influence provider-patient dynamics, even as patients increasingly arrive with their own data and questions. In this context, the precision movement must avoid becoming yet another layer of abstraction. Algorithms can assist, but only people can reassure.

Patient-centered design must now lead the next wave of transformation. This means interfaces that explain rather than obscure, systems that empower clinicians to connect, and care plans that reflect not only genomics but also values, preferences, and lived experience. Our guides, Michael, Jennifer, and Emily, benefited from the incorporation of unique data about them as individuals, with wearable technology and pharmacogenomic testing, respectively. It also means reimagining roles: health coaches, genetic counselors, and data concierges who translate complex information into compassionate action, like the way Amy benefited with Big Data insights about environmental factors in pain management, and Susan's physician identified an AI-guided treatment for her advanced-stage cancer. The goal is not just precise care but care that is deeply personal because it is built on trust.

Interprofessional Teams and the New Model

The traditional structure of care delivery—that is, physician-led, specialist-referred, and episodic—was never built to manage complexity at the scale required by precision medicine. One clinician, no matter how expert, cannot simultaneously interpret genomic data, manage behavior change, validate AI outputs, and coordinate longitudinal care. Yet that is often the expectation. The result is burnout, decision fatigue, and underutilized innovation. It is not a failure of clinicians; it is a failure of the model.

The new model demands interprofessional teams organized not just by specialty (when surgical/treatment experience with an anatomical system is necessary) but by stage of care: forologists, acutologists, and chronicists, as mentioned previously. These roles will likely redefine primary care, refining the generalist perspective with more precise functions. Pharmacists, genetic counselors, and nurse navigators must be integrated, not referred, as core collaborators.

Team-based care is not just a coordination problem; it is an identity shift. It redefines the clinician not as a lone expert but rather as a node in a dynamic network. It also requires systems to support trust, shared goals, and transparent communication, qualities too often neglected in siloed environments. In the precision era, the team is the unit of care, and the patient is its center. That is not a dilution of medicine. It is its evolution.

How Precision Medicine Will Shape Personalized Healthcare

Precision medicine is laying the foundation for a truly personal health-care ecosystem—one where care decisions emerge not from averages but instead from the unique interplay of a person's biology, behavior, and environment. This is more than individualized treatment; it's the creation of a living, adaptive health environment that grows with the patient. Precision tools, from wearables and genomics to digital therapeutics and AI, serve as both microscope and compass, illuminating what makes each patient distinct and guiding care accordingly.

In this ecosystem, care is not something done to the patient but something shaped with them. A cancer diagnosis, for example, no longer triggers a standardized sequence of chemotherapy options; instead, as we saw with Susan's case, it initiates a genomic analysis, incorporates a side-effect-risk profile, and establishes a shared decision-making process around evidence-based options tailored to the patient's DNA and goals. Over time, predictive analytics and real-world feedback continuously refine this model, creating a dynamic loop between lived experience and clinical insight. This is personalized healthcare not as a marketing term but as a functional reality—and precision medicine is the engine making it possible.

Empowered Patients, Responsive Systems

At the heart of personalized healthcare is empowerment: the idea that patients are not just recipients of care but collaborators in it. Precision medicine offers the tools to make that empowerment real. When patients understand how their wearable data reflects behavior change, how their genes affect medication response, or how AI models predict risk based on their specific profile, they become more than informed—they become engaged. This engagement transforms healthcare from compliance to collaboration. No longer bound by generic care plans or "one-size-fits-all" advice, patients increasingly co-create their treatment strategies with clinicians. Responsive systems, enabled by interoperable data, transparent AI, and value-aligned care teams, adapt as patients' lives evolve. A wearable sleep tracker might prompt a change in medication timing; a shift in stress levels might trigger a virtual behavioral health check-in. In each case, the system listens and adjusts.

This patient-centered responsiveness is not just a technical advancement—it's a human one. It respects the complexity of each life, giving patients the dignity of relevance and voice. In doing so, it restores a sense of agency in an arena that often feels opaque and disempowering. Precision medicine brings both clarity and possibility, and with them, a new era of partnership between people and their care. This approach of precision medicine transforms the structure of care into ongoing, data-driven relationships. Now, real-time data from sensors, smart devices, and genomic updates enables continuous connection. The result is a care relationship that is ambient, proactive, and profoundly personal.

Imagine a hypertensive patient whose wearable detects subtle changes in heart rate variability and sleep quality, prompting an early adjustment in medication or lifestyle. Or a patient recovering from surgery, whose wound-healing sensor and mobility tracker reduce the need for in-person checkups while still ensuring safety and recovery. These examples are not future concepts; they're becoming the standard in systems adopting precision tools.

Importantly, this always-on model doesn't eliminate the need for in-person care; it enhances it. Physical visits are enriched by longitudinal data, and conversations shift from "What happened since I last saw you?" to "Here's what's changing, and why." The relationship deepens as technology fills the gaps between visits, supporting patients with both consistency and context. In this model, care isn't delivered in bursts; it's woven into the patient's life, creating a partnership that adapts in real time.

Precision Across the Lifespan

As this new perspective makes clear, personalized healthcare is not an intermittent reaction; it's a longitudinal journey. Precision medicine brings this into sharp focus, enabling tailored interventions across every stage of life. From newborn genomic screening that identifies rare conditions before symptoms emerge, to pharmacogenomic guidance that shapes adolescence and adulthood, to AI-informed frailty monitoring in older age, the precision approach doesn't fade—it evolves.

For example, a child with a genetic predisposition to a certain metabolic disorder can begin a lifestyle intervention plan in kindergarten, co-developed with pediatricians and nutritionists. In midlife, that same individual might receive AI-supported cancer risk assessments shaped by decades of personal and family history. Later in life, passive sensors might help caregivers and clinicians detect early cognitive decline before it becomes disabling. Each stage is defined not by a protocol but by a conversation—guided by data, shaped by context, and centered on what matters most at that moment.

This across-the-lifespan model is more than continuity; it is intentional design. It integrates the changing biology, priorities, and vulnerabilities of each life phase, and it does so without demanding that patients start over with each new diagnosis. Precision medicine creates the thread that ties these stages together, offering continuity not only of care but also of identity and purpose. And although precision medicine is rooted in individual variation, its benefits extend well beyond the individual. Personalized healthcare, when deployed effectively, can reduce waste, improve outcomes, and enhance workforce productivity, delivering economic returns as well as clinical ones. Chronic conditions, which drive the majority of healthcare costs, often stem from poor alignment between treatment and personal context. Precision medicine corrects that misalignment.

Consider medication adherence: as we learned in Chapter 5, the preponderance of all medications are not taken as prescribed, leading to preventable complications and hospitalizations. With pharmacogenomics and personalized coaching, we can tailor prescriptions and identify the behavioral barriers that keep patients from following through. The result is not only better outcomes but also fewer wasted prescriptions, fewer ER visits, and lower employee absenteeism.

At a broader level, personalization reduces the hidden costs of mismatch—whether between care and culture, treatment and biology, or recommendations and reality. By making care more efficient, effective, and satisfying, it has the potential to bend the cost curve while raising the standard of care. Precision medicine is not just a clinical breakthrough; it is a system-level improvement that, if scaled wisely, can help build a more sustainable and humane healthcare economy.

Toward a New Definition of Healthcare—and Health

One of the deepest fears about technology in healthcare is that it will make care more impersonal. But when precision medicine is deployed with intention, it does the opposite. By eliminating the guesswork, the repetition, and the feeling of being misunderstood, it clears space for connection. When a physician already knows which medication is likely to work and why, it frees up time to ask, "How are you really doing?"

In a precision-enabled system, empathy is not sacrificed to data; it's augmented by it. AI tools that draft notes can let clinicians look patients in the eye again. Predictive algorithms that flag rising anxiety can trigger a call before a patient reaches a breaking point. Genomic insights can help explain suffering that was once dismissed. These are not cold efficiencies; they are acts of recognition.

Empathy, in this future, is not measured by bedside manner alone but instead by relevance, understanding, and responsiveness. It is about being seen and known, not just statistically, but personally. Precision medicine provides the map. Personalized healthcare makes the journey humane. Together, they bring us closer not just to healing, but to being heard.

> Precision medicine doesn't just change how we treat disease; it changes how we define health.

The amazing conclusion of this paradigm shift is this: precision medicine doesn't just change how we treat disease; it changes how we define health. In a system built around personalization, health becomes less about achieving arbitrary thresholds and more about aligning with each person's goals, capacities, and context. It's a shift from treating pathology to cultivating well-being, on terms that are meaningful to the individual.

In this model, a healthy life for a retired schoolteacher managing arthritis may involve movement goals tailored to her baseline joint function, a medication plan optimized to avoid sedation, and a sleep schedule shaped by circadian rhythm data. For a young adult with a genetic predisposition to anxiety, it might mean daily heart rate variability tracking, early behavioral support, and community engagement plans. Each person's health is defined by what enables them to thrive, not just survive.

This redefinition is not idealistic; it is deeply practical. It enables clinicians to deliver care that is both clinically effective and personally meaningful. It empowers patients to pursue health in a way that reflects their lives, not someone else's average. Precision medicine unlocks the tools; personalized healthcare gives them purpose. Together, they mark a new chapter—one in which the future of medicine is not just more advanced but more human.

Notes

1. Unlearn.AI, "Digital Twin Generators," (2025), `https://www.unlearn.ai/digital-twin-generators`.

2. Tempus, "Tempus Announces National Launch of Olivia, Its AI-Enabled Personal Health Concierge App for Patients," January 21 (2025), `https://www.tempus.com/news/pr/tempus-announces-national-launch-of-olivia-its-ai-enabled-personal-health-concierge-app-for-patients`.

3. Abbott, "3 Ways Biowearables Will Be Advanced Health Trackers," April 26 (2023), `https://www.abbott.com/corpnewsroom/products-and-innovation/3-ways-biowearables-will-be-advanced-health-trackers.html`.

4. Apple, "Groundbreaking Health Features Available Today on Apple Watch and AirPods Pro 2," March 26 (2025), `https://www.apple.com/my/newsroom/2025/03/groundbreaking-health-features-available-today-on-apple-watch-and-airpods-pro-2`.

   Reuters, "Apple to Bring Satellite Communications to Smartwatch Next Year, Bloomberg News Reports," December 10 (2024), `https://www.reuters.com/technology/apple-bring-satellite-communications-smartwatch-next-year-bloomberg-news-reports-2024-12-10`.

   Balasubramanian, Sai, "Here's What We Know about Project Mulberry—Apple's Secret Initiative to Redesign Healthcare," April 16 (2025), `https://www.forbes.com/sites/saibala/2025/04/16/heres-what-we-know-about-project-mulberry--apples-secret-initiative-to-redesign-healthcare`.

   Apple, "New Holistic Apple Health Study Launches Today in the Research App," February 12 (2025), `https://www.apple.com/newsroom/2025/02/new-holistic-apple-health-study-launches-today-in-the-research-app`.

5. Validic, "Validic Advances Personalized Patient Care by Integrating Wearable Data in EHR Workflow," January 8 (2025), `https://www.validic.com/resources/news/validic-advances-personalized-patient-care-by-integrating-wearable-data-in-ehr-workflow`.

6. Mayo Clinic Center for Individualized Medicine, "RIGHT 10K," May 2 (2025), `https://www.mayo.edu/research/centers-programs/center-individualized-medicine/research/clinical-studies/right-10k`.

Archiquette, John, "Pharmacogenomics and How the VA Is Improving the Efficacy of Medicine through DNA," October 21 (2024), https://www.va.gov/southern-nevada-health-care/stories/pharmacogenomics-and-how-the-va-is-improving-the-efficacy-of-medicine-through-dna.

7. Genomind, "Pharmacogenetic Testing," May 2 (2025), https://genomind.com/solutions/pharmacogenetic-testing.

    Gene By Gene, "Pharmacogenomic Testing," May 2 (2025), https://www.genebygene.com/service/pharmacogenomic-testing.

8. Greene, Jan, "Machine Learning Analysis of Sepsis Illustrates Condition's Complexity," July 17 (2019), https://divisionofresearch.kaiserpermanente.org/machine-learning-analysis-of-sepsis.

9. Rehan, Hassan, "Enhancing Early Detection and Management of Chronic Diseases with AI-Driven Predictive Analytics on Healthcare Cloud Platforms," *Journal of AI-Assisted Scientific Discovery* 4, no. 2 (2024): 1-38.

    Ma, Lijun, and Tibble, Molly, "Primary Care Asthma Attack Prediction Models for Adults: A Systematic Review of Reported Methodologies and Outcomes," *Journal of Asthma and Allergy* 17 (2024): 181-194.

10. Dahdah, Robert, "A Year of DAX Copilot: Healthcare Innovation That Refocuses on the Clinician-Patient Connection," September 26 (2024), https://blogs.microsoft.com/blog/2024/09/26/a-year-of-dax-copilot-healthcare-innovation-that-refocuses-on-the-clinician-patient-connection.

11. PathAI, "Pathology Transformed," May 5 (2025), https://www.pathai.com.

    Aidoc, "AI Solutions for Radiology," May 5 (2025), https://www.aidoc.com.

12. Neuralink, "Neuralink: Developing the Future of Brain-Computer Interfaces," May 5 (2025), https://neuralink.com.

13. Office of the National Coordinator for Health Information Technology, "What Is FHIR?" August (2019), https://www.healthit.gov/sites/default/files/2019-08/ONCFHIRFSWhatIsFHIR.pdf.

14. Raths, David, "TEFCA QHIN-to-QHIN FHIR Exchange to Be Piloted in 2025," January 20 (2024), `https://www.hcinnovationgroup.com/interoperability-hie/trusted-exchange-framework-and-common-agreement-tefca/news/53083106/tefca-qhin-to-qhin-fhir-exchange-to-be-piloted-in-2025`.

15. PointClickCare, "Connected Health for Senior Care," May 5 (2025), `https://pointclickcare.com`.

16. Epic Systems Corporation, "Epic Cosmos," May 5 (2025), `https://cosmos.epic.com`.

    Amazon Web Services, "Amazon HealthLake," May 5 (2025), `https://aws.amazon.com/healthlake`.

17. Omada Health, "Omada Health: Digital Care That Changes Lives," May 5 (2025), `https://www.omadahealth.com`.

18. Current Health, "Remote Care Management Platform," May 5 (2025), `https://www.currenthealth.com`.

Conclusion: Making Medicine Personal

In the traditional practice of medicine, care often begins with averages. Clinicians rely on population-based data, clinical guidelines, and one-size-fits-all protocols to determine a course of action. Although this approach has enabled broad public health advances, it treats variability as noise to be minimized rather than insight to be harnessed.

The era of real-time, individualized healthcare, however, is here. Enabled by technological breakthroughs across four key drivers—wearable devices, genomics, Big Data analytics, and artificial intelligence—and enhanced by burden-reducing integration, precision medicine is reshaping the clinical foundation of healthcare.

The Personal Healthcare Journey

Every health journey is personal, but until now, medicine has rarely treated it that way. Today's technology gives us the tools to listen to each person's body, biology, and lived experience in real time. The following are the breakthroughs making that possible, and how they turn hope—and data—into truly personal care.

Wearables: Real-Time Data Inputs

Wearable devices represent a remarkable paradigm shift in how data enters the healthcare ecosystem. Rather than depending on sporadic checkups or self-reported symptoms, clinicians now receive biometric data directly (and continuously, if helpful) from the patient's body. From heart rate variability to sleep patterns, oxygen saturation to glucose levels, wearable devices deliver a stream of real-time inputs that can inform treatment, alert users and providers to early signs of deterioration, and support behavioral interventions. These tools are no longer reserved for fitness enthusiasts; they are evolving into clinical-grade devices capable of flagging atrial fibrillation, predicting insulin spikes, and improving management of chronic diseases.

From a process standpoint, wearables extend the clinical boundary beyond the exam room. They introduce a feedback loop that empowers patients and provides clinicians with context-rich, longitudinal datasets. This transition repositions care from episodic to continuous, from reactive to proactive. Most importantly, perhaps, they involve the patient in the course of care.

Genomics: Patient-Specific Inputs

Beyond wearables, which provide dynamic external data, genomics offers a static but foundational layer of the biological blueprint. Genomic sequencing and pharmacogenomic testing allow providers to understand how a patient's genes influence disease risk, drug metabolism, and therapeutic response. This is a powerful clinical advancement. With a simple genetic test, physicians can tailor medications to align with an individual's metabolic pathways, drastically reducing trial-and-error prescribing.

Genomics is also unlocking new paths for early diagnosis and targeted intervention. From identifying cancer susceptibility to pinpointing rare disease variants, genomics serves as a roadmap that informs nearly every subsequent decision in the care continuum. Clinically, genomics is shifting the decision tree: not what is likely to work for the average person but what is most likely to work for this person, based on their genetic profile.

Big Data: Macro-Intelligence

As genomic and wearable data grow in volume and complexity, however, they must be placed within a broader context to gain meaning. Big Data analytics enables this by aggregating, de-identifying, and analyzing millions of data points across clinical records, social determinants of health, lab results, genomic sequences, and wearable outputs. At scale, this allows for pattern detection, risk prediction, and stratified care planning that's simply not possible in $n=1$ reactive care.

Big Data allows healthcare to move from treating disease to anticipating it. Even if a patient may not meet current clinical criteria for diabetes, by comparing their lifestyle, biometrics, and family history with massive datasets, clinicians can identify precursors and intervene years earlier than traditional methods would suggest. Big Data also supports precision at the population level, identifying treatment efficacy across subgroups, uncovering rare associations, and continuously updating clinical decision pathways based on real-world outcomes.

Big Data platforms form the backbone of next-generation clinical intelligence systems, integrating diverse data sources and feeding into decision support tools. This transforms anecdotal care into data-driven practice.

Artificial Intelligence: The Engine of Integration

At the endpoint of the precision medicine ecosystem is artificial intelligence—the integrative force that transforms raw data into actionable insight. AI algorithms can process the complexity of genomics, wearables, and population data far faster and more accurately than human cognition allows. Machine learning models detect patterns, anomalies, and correlations across multimodal data, enabling everything from early cancer detection to real-time optimization of care plans.

But AI is neither a single tool nor an endpoint—it is a spectrum of technologies, existing in a circle of continuous improvement: natural language processing to extract meaning from unstructured clinical notes, computer vision to interpret imaging, predictive analytics to model disease progression, and recommendation engines that guide precision prescribing. Its power lies in orchestration—synthesizing dynamic (wearables), static (genomic), and contextual (Big Data) inputs to help clinicians make better, faster, and more personalized decisions.

From a process perspective, AI reshapes the clinical workflow. Rather than triaging based solely on symptoms or guidelines, AI-driven systems can stratify risk, recommend diagnostics, and surface best-fit treatments automatically. These systems do not replace human judgment; they augment it with depth, speed, and scope previously not possible.

Toward a New Clinical Model

Together, these four drivers form a continuous system: wearables collect real-time data; genomics anchors the data in a personalized framework; Big Data contextualizes it; and AI translates it into action. This architecture represents a procedural leap from generalized care toward dynamic, individualized pathways.

In the old model of traditional medicine, care was driven by what worked for most people. In the precision medicine model, care is driven by what works for you. As this new system takes root, clinical processes will need to adapt, moving from static protocols to adaptive pathways, from patient visits to patient streams, and from retrospective analysis to real-time response.

This is no longer the medicine of the future. It is here, now—an operational, procedural, and scientific framework that elevates both outcomes and experience. And although technology enables it, the goal remains profoundly human: to make medicine personal.

Common Themes from Our Guides: What We Can Learn

The technologies behind precision medicine, wearables, genomics, Big Data, and AI are abstractions; however, their power is revealed through people. The stories of Michael and Jennifer, Emily, Amy, and Susan bring to life what clinical theory alone cannot: the human transformation that occurs when medicine becomes personal.

Michael and Jennifer: A New Feedback Loop

Michael and Jennifer's journey began not in a clinic but at home—with two smartwatches and a quiet hope. They had spent years struggling with their health and weight, cycling through diets and routines without meaningful results. But when wearable devices began providing real-time feedback on sleep quality, activity levels, heart rate, and glucose variability, they encountered something new: visibility. Patterns emerged, energy levels correlated with food choices, and step counts told a story about effort, not just outcome.

The continuous stream of personal data didn't just inform their choices; it changed them. As the watches nudged them to move, rest, hydrate, or adjust their meals, they began to trust the data over their frustrations. Michael's blood pressure and glucose readings revealed daily fluctuations that allowed him to modify behaviors in real time. Jennifer's sleep tracker showed she was barely entering deep sleep, prompting lifestyle changes that improved both rest and recovery. Through their devices, they created a feedback loop between insight and action. Wearables helped them reclaim their bodies, their partnership, and a sense of control.

Emily: The Genomic Breakthrough

Emily's story was different but born of the same frustration. Diagnosed with anxiety and depression, for years she had tried multiple medications with little relief and many side effects. Each failed attempt added another layer of doubt and hopelessness. Then came a turning point: a pharmacogenomic test that decoded her personal genetic profile.

The results were eye-opening. Certain antidepressants she had been prescribed were metabolized too quickly to be effective; others posed heightened risks of adverse reactions. With her genomic map in hand, her provider recommended a medication with a high likelihood of therapeutic success—and this time, it worked. Within weeks, Emily began to feel clarity and emotional steadiness she hadn't known in years. Genomics didn't just help choose the right medication. It validated her experience, proving that her previous struggles were not imaginary or a failure of willpower, but mismatches between her biology and the treatments she was given. It was the science of self, used to heal.

Amy: When Big Data Sees What Others Miss

Amy was different. She had lived for years with chronic pain that defied diagnosis, each appointment bringing new tests, new referrals, and the same disappointing conclusion: inconclusive. Her pain was real, but its cause remained hidden—until a precision medicine program integrated her case into a Big Data platform.

By comparing her records, symptoms, and biometrics with thousands of others, a pattern emerged: an opaque but increasingly recognized set of neurological stressors. Suddenly, the once-invisible circumstances had an identity. Treatment began, based not on assumption but instead on statistical correlation refined through machine pattern recognition. Amy's story is a testament to how Big Data can unearth clarity from complexity. It gave her a path forward: one rooted not in generic advice but in informed recognition.

Susan: AI at the Limits of Hope

In our last story, Susan's cancer diagnosis arrived late, with few options under standard protocols. But her oncologist was part of a research initiative using AI to match patients with potential therapies based on tumor genomics, treatment histories, and outcomes from similar cases worldwide.

Through this AI-driven model, Susan was matched with a targeted therapy that wouldn't have been considered in prior care. The algorithm recognized a genomic mutation that made her a candidate for a specific immunotherapy, based on a small but significant cluster of prior cases. The result: her cancer regressed, and her quality of life improved dramatically.

For Susan, AI wasn't some cold machine—it was a lifeline. It extended her treatment options beyond what even seasoned specialists could see. Where others saw the end of the road, AI uncovered the hidden path.

From Personal Data to Personalized Care

Across these stories, we see precision medicine not as a futuristic concept but as a practical, life-altering system. Each technology served a different role: wearables provided real-time feedback, genomics explained therapeutic mismatches, Big Data revealed population-level patterns with individual relevance, and AI brought synthesis and insight across complexity.

These are not isolated anecdotes. They reflect a replicable model of care that is data-driven but deeply human. What unites these journeys is not just better outcomes but restored agency. Michael and Jennifer became partners in their health. Emily found not just a treatment but an explanation. Amy was finally heard, and Susan was seen.

Their experiences confirm a central truth: when medicine recognizes the uniqueness of the person, it becomes exponentially more effective. The clinical tools behind precision medicine are powerful, but they are only as meaningful as the lives they improve. In that sense, the stories of our guides are not the exception—they are the new standard we must pursue.

How You Can Explore Precision Medicine for Yourself

The north star of our journey in this book is that the promise of precision medicine isn't reserved for specialists or researchers. It's not limited to major hospitals or future generations. It's available, right now, to anyone willing to ask better questions, explore new tools, and advocate for more personal care. Whether you're a patient, a caregiver, or simply someone looking to take charge of your health, this section offers a practical guide to getting started.

At Home and In Your Daily Life

The steps in making precision medicine work for you are simple, yet extraordinarily effective.

Start Small, Start Smart

The entry point to precision medicine doesn't require a lab coat or a geneticist. It might begin with something as simple as a wearable device. Fitness trackers, smartwatches, and mobile health apps can help you observe trends in sleep, activity, heart rate, and glucose levels. These aren't just curiosities—they're your daily data, and they offer valuable insights when used consistently.

Pay attention to patterns. What happens when you eat late, skip sleep, or exercise regularly? Your body is already giving you feedback. Precision medicine starts by listening to it.

Explore Tools That Empower, Not Just Inform

From symptom trackers to connected blood pressure cuffs to AI-powered apps that monitor mental health or medication adherence, there are dozens of precision-focused tools that don't require a prescription. Use them to enhance awareness, not to chase perfection. The goal isn't to live by the numbers but to be guided by them.

Many platforms now integrate wearable and lab data with recommendations tailored to your age, goals, and conditions. Some even offer virtual coaching, medication reminders, and mood or stress tracking. Used wisely, these tools can create a comprehensive, day-to-day portrait of your health journey—informing not just what you do but when and how you do it.

With Your Family and Friends

Healing is easier when you're not alone. Online communities focused on precision medicine, chronic illness management, biohacking, and condition-specific support can be invaluable. Not only do they offer insight into what others are trying, but they often surface new tools, providers, and research that you might not find through traditional channels.

In sharing your story, you also help others. Precision medicine thrives on diversity—genetically, behaviorally, and demographically. The more stories are shared, the more research evolves, and the better the care becomes for everyone.

With Your Healthcare Provider

In this new environment of precision medicine, your physician as well as your pharmacist, nurse navigator, and care manager are your support team. Once reserved for elite athletes, the services at their disposal now make you part of that elite team.

Understand the Power of Your DNA

Genetic testing, particularly pharmacogenomic testing, can help explain why certain medications work well for some people and not others. If you've ever had a poor response to a drug or cycled through antidepressants, blood pressure medications, or pain relievers without consistent relief, you may benefit from a test that evaluates how your body metabolizes common drugs.

These tests are becoming more accessible through both clinicians and direct-to-consumer platforms. Increasingly, insurance plans and Medicare Advantage programs cover them. If you're on long-term medications or managing a chronic condition, bring it up with your provider: "Would pharmacogenomic testing help personalize my treatment?" The answer can open a new chapter in your care.

Connect the Dots with a Care Team That Listens

Data without understanding is noise. Precision medicine works best when your healthcare providers know how to interpret and act on the insights you bring them. This means finding physicians, nurse practitioners, or care teams who value patient-specific inputs, use digital tools, and are open to collaboration.

You can help this process by organizing your information. In the immortal words of Jerry Maguire, "Help me help you"—bring your wearable data to appointments.[1] Share your symptoms in patterns, not episodes: "This always happens after a poor night's sleep," or "My glucose spikes an hour after this type of meal." If you've had genetic testing, bring the results. You are not just reporting symptoms; you are becoming a key player in your own healing.

Ask Better Questions

Finally, start every clinical encounter with one powerful idea: "What would make this more personal?" Instead of asking, "What's the standard treatment?" ask, "What would work best for me, given what we know about my body, my habits, and my goals?"

You may not always get a perfect answer, but you will change the conversation. You will introduce the idea that you are not an average patient—you are a specific person seeking care tailored to your reality.

Make Progress, Not Perfection

Above all, it's important to remember that precision medicine isn't about micromanaging every calorie, step, or data point; it's about working smarter, not harder. You're not chasing control, you're building awareness. Real improvement often begins with one small, sustainable change: more movement, better sleep, a medication change informed by a genetic test. And often, precision medicine offers peace of mind. It confirms that what you've felt was real, and that you weren't wrong. That better options exist.

You don't need to be a scientist to engage with this new paradigm of healthcare. You just need to be curious, proactive, and open to learning. Precision medicine becomes powerful when patients demand it, use it, and share their success. Every time you choose a more personal approach, you help move the system in the right direction.

Personalized care is not a promise for the future. It is a choice you can make—today.

Call to Action: A Roadmap for Leaders, Investors, and Policymakers

The transformation of medicine is not only a clinical imperative—it is a leadership challenge. The tools of precision medicine are available, proven, and increasingly affordable. Yet their adoption lags, not because of technological limits but due to inertia in systems, incentives, and expectations. If we are to realize the full promise of personalized healthcare, we need visionary action from every corner of the ecosystem. The roadmap forward is not abstract. It is clear, achievable, and urgently needed.

For Healthcare Leaders and Providers: Rethink Delivery

Healthcare executives, hospital administrators, and clinical leaders must ask: are we organized around the average or the individual? Precision medicine requires moving beyond episodic care and fragmented specialties to embrace continuous, longitudinal, and interdisciplinary models.

Clinical workflows must incorporate wearables, genomics, and AI tools as core inputs, not add-ons.

This can be accomplished by

- Investing in care models that integrate remote monitoring, pharmacogenomic decision support, and predictive analytics
- Training providers not only in how to use these tools but also in how to engage patients as partners in a shared data journey
- Breaking down silos between primary care, specialty care, and digital health, and building delivery paths that are longitudinal and oriented on phase of care, not anatomical system or complexity
- Making precision medicine the default, not the exception

For Payers and Insurers: Shift the Incentives

Traditional fee-for-service models reward volume, not outcomes. But precision medicine thrives on insight and impact: preventing complications, reducing medication misfires, and improving quality of life. Insurers, self-insured employers, and other payers must revise their frameworks to recognize the long-term economic value of personalized approaches.

This can be accomplished by

- Covering pharmacogenomic testing not just for cancer or rare disease but for common conditions like depression, pain, and cardiovascular care
- Reimbursing providers for time spent interpreting wearable and genetic data
- Supporting value-based arrangements that incentivize outcome-driven innovation

The bottom line: the most expensive care is ineffective care. Precision medicine helps avoid it.

For Policymakers and Public Institutions: Democratize Access

The benefits of precision medicine must not be confined to urban centers and academic medical institutions. Policymakers have a critical role to play in ensuring equity. That begins with something as simple as supporting broadband access and digital literacy, especially in rural and underserved communities.

This can be accomplished by

- Funding pilot programs that bring precision diagnostics and AI-based clinical tools into community clinics
- Expanding public insurance coverage to include wearable monitoring and genetic testing for chronic conditions
- Creating public health initiatives that incorporate real-time data, not just retrospective metrics
- Ensuring that privacy protections evolve alongside data capabilities

The role of policy here is natural and obvious: precision medicine should not deepen disparities; it should help close them. This requires intentional design, inclusive data, and minimally essential regulation that fosters both innovation and equity.

For Investors and Innovators: Build the Bridges

The next decade of healthcare value creation will come not from marginal gains within the old system but from enabling the new one. Investors and entrepreneurs should focus on companies and platforms that integrate the four drivers—wearables, genomics, Big Data, and AI—into actionable clinical workflows.

The most valuable innovations will be those that lower the barrier to entry for personalized care. That means user-friendly interfaces, interoperable data platforms, consumer-facing diagnostics, and digital tools that empower patients while informing clinicians. Capital should flow to solutions that serve both individual precision and system-level efficiency.

For Medical Educators and Researchers: Train for the Future

Similar to the recommendations of the Flexner Report over a century ago, medical education must evolve to reflect the reality that medicine is no longer static.[2] In the third decade of the 21st century, new clinicians must be taught how to work with genomic data, interpret real-time patient monitoring, and collaborate with AI decision tools. Research institutions must prioritize the integration of these technologies in clinical trials and outcomes research.

For academics, this means several things. In developing curricula, educators should emphasize patient variability, data literacy, and collaborative care. In performing innovative research, investigators should ensure that trials include diverse populations, real-world data inputs, and adaptive methodologies. The future of medicine is agile, personalized, and interdisciplinary, and the training must reflect that.

A Shared Responsibility

Finally, however enthusiastic, no single group of stakeholders can implement precision medicine at scale. But collectively, we can create the infrastructure, incentives, and culture required to make it universal. The stories in this book, of authentic people experiencing realizable transformations, are not exceptions. They are glimpses of what is possible.

For precision medicine to achieve its full potential—the true personalization of healthcare—paradigms must change across the spectrum of healthcare involvement. Leaders must lead. Investors must invest. Policymakers must protect and enable. Providers must adapt and adopt. And patients must demand care that sees them not as averages but as individuals. In this manner, the collective efforts produce more value collaboratively, not less competitively.

> **Leaders must lead. Investors must invest. Policymakers must protect and enable. Providers must adapt and adopt. And patients must demand care that sees them not as averages but as individuals.**

Unlike most other products and services, healthcare is not a zero-sum game. Healthcare delivered via traditional medicine is actually a negative-sum enterprise: delaying appropriate care benefits the financial party less than it costs the patient encountering the suboptimal outcome. Precision medicine, on the other hand, is a positive-sum activity: by embarking on the course of action, all parties are better off. Precision medicine is not a luxury; it is a moral and economic imperative.

We have the tools; all we need is the will. And the moment is now—the future of healthcare is personal; let's build it.

Notes

1. *Jerry Maguire*, directed by Cameron Crowe, TriStar Pictures (1996).
2. Flexner, "Medical Education in the United States and Canada."

Acknowledgments

This book wouldn't be without some remarkable people.

To my family—thank you for grounding the patient stories in reality. Your insights, candor, and experiences helped breathe authenticity into the examples that carry these pages. The vignettes are more vivid because of you.

To Regina Herzlinger and Richard Dawkins, whose works in adjacent fields defined clear yet accessible ways to tackle complex ideas—your examples were a steady guidepost. And to Alvin Toffler and Elon Musk—their brief intersections in my life deeply influenced my desire to visualize and describe the future, Toffler through his work and legacy, and Musk through current-day vision and provocation.

I am immensely grateful to the exceptional team at Wiley. To Kenyon Brown, my insightful Senior Editor, for championing the work from the very beginning. To Navin Vijayakumar, Managing Editor, for keeping every part of the process moving forward with calm command. To Gus Miklos, Project Editor, for refining tone and narrative flow with persistent care. And to Tiffany Taylor, Saravanan Dakshinamurthy, and Anjali Godiyal—thank you for your behind-the-scenes brilliance in content refinement and permissions that made this book stronger and more appealing.

To everyone who offered feedback, encouragement, or challenge along the way—thank you. Your support helped shape this book and sharpen its purpose.

About the Author

James Wallace is a healthcare strategist, researcher, and chief executive with deep expertise in personalized medicine. A graduate of the United States Military Academy, he holds an MBA from Harvard Business School and a Doctor of Business Administration degree from the University of South Florida.

Over a career spanning innovation, delivery, and policy in multiple disciplines, Jim has advised providers, investors, and policymakers on the real-world application of artificial intelligence, emotive–cognitive systems, and precision medicine.

Jim is a frequent speaker at national and international forums shaping the future of healthcare. When not delivering on the promise of the future, he enjoys hiking and camping in New Hampshire and Arizona—and eagerly looks forward to joining the first Starship to Mars.

Index